MATH MADE A BIT EASIER LESSON PLANS

Also by Larry Zafran

WEIGHT LOSS MADE A BIT EASIER:
Realistic and Practical Advice for Healthy Eating and Exercise

A REALISTIC EATING AND EXERCISE RECORD BOOK:
A Six-Month Weight Loss Log and Journal
for Dedicated Individuals

THE COMPLETE MUSIC PRACTICE RECORD BOOK:
A Six-Month Log and Journal for Dedicated Students

AMERICA'S (MATH) EDUCATION CRISIS:
Why We Have It and Why We Can('t) Fix It

THE REGIFTABLE GIFT BOOK:
The Gift That Keeps On Regiving

MATH MADE A BIT EASIER:
Basic Math Explained in Plain English

MATH MADE A BIT EASIER WORKBOOK:
Practice Exercises, Self-Tests, and Review

BASIC ALGEBRA AND GEOMETRY MADE A BIT EASIER:
Concepts Explained in Plain English,
Practice Exercises, and Self-Tests

BASIC ALGEBRA & GEOMETRY M.A.B.E. LESSON PLANS:
A Guide for Tutors, Parents, and Homeschoolers

www.LarryZafran.com

MATH MADE A BIT EASIER LESSON PLANS

A Guide for Tutors, Parents, and Homeschoolers

LARRY ZAFRAN

Self-published by Author

MATH MADE A BIT EASIER LESSON PLANS:
A Guide for Tutors, Parents, and Homeschoolers

Copyright © 2010 by Larry Zafran
Self-published by Author

All rights reserved. No part of this book may be reproduced or transmitted in any manner whatsoever without written permission except in the case of brief quotations embodied in critical articles and reviews.

Book design by Larry Zafran

Printed in the United States of America
First Edition printing January 2010
First Edition revised May 2011

ISBN-10: 1-4499-9709-0
ISBN-13: 978-1-44-999709-0

Please visit the companion website below for additional information, to ask questions about the material, to offer feedback, or to contact the author for any purpose.

www.MathWithLarry.com

CONTENTS

CHAPTER ZERO 7
Introduction

CHAPTER ONE 13
How to Be an Effective Math Instructor

CHAPTER TWO 23
Lesson Plans for "The Foundation of Math:
    Basic Skills in Arithmetic"

CHAPTER THREE 37
Lesson Plans for "Basic Math Topics
    and Operations"

CHAPTER FOUR 49
Lesson Plans for "Working with
    Negative Numbers"

CHAPTER FIVE 59
Lesson Plans for "Basic Operations
    with Fractions (+, −, ×, ÷)"

CHAPTER SIX 65
Lesson Plans for "More About Fractions"

CHAPTER SEVEN 75
Lesson Plans for "Other Topics in Fractions"

CHAPTER EIGHT 81
Lesson Plans for "The Metric System, Unit Conversion, Proportions, Rates, Ratios, and Scale"

CHAPTER NINE 91
Lesson Plans for "Working with Decimals"

CHAPTER NINE AND FIVE-TENTHS 99
Lesson Plans for "More Topics in Decimals"

CHAPTER TEN 107
Lesson Plans for "Working with Percents"

CHAPTER ELEVEN 119
Lesson Plans for "Basic Probability and Statistics"

CHAPTER TWELVE 129
How to Teach Meditation to Increase Mindfulness and Reduce Anxiety

About the Author & Companion Website 137

CHAPTER ZERO

INTRODUCTION

ABOUT THE *MATH MADE A BIT EASIER* SERIES

This is the third book in the self-published *Math Made a Bit Easier* series which will be comprised of at least nine books. The goal of the series is to explain math "in plain English" as noted in the subtitle of the first book.

The first book in the series covers basic math which is the foundation of all later math that a student will study. It is essential that the first book not be skipped, and that it is thoroughly mastered before taking on later math. The book is available for free reading in its entirety on *Google Books* for those who cannot or choose not to purchase it.

The second book in the series is a companion workbook of review, practice exercises, and self-tests which corresponds directly to the first book. This book makes frequent reference to the workbook, so it too is available for free reading on the Internet for those who are unable or do not wish to buy it.

The next set of three books in the series will cover basic algebra and geometry. With that material mastered, the student should be well-prepared for standardized exams such as the SAT or GED, with the exception of a few more advanced topics. The next set of three books will cover more advanced topics in algebra and geometry. A tenth book in the series will discuss the American math education system.

In addition to providing math instruction, the series also attempts to explain the truth about why students struggle with math, and what can be done to remedy the situation. As a totally independent and self-published author, I am able to write with such candidness.

Unlike many commercial math books, the series does not imply that learning math is fast, fun, or easy. It requires time and effort on the part of the student. It also requires that the student be able to remain humble as s/he uncovers and fills in all of his/her math gaps, and that s/he makes time to do so.

THE PURPOSE AND TARGET AUDIENCE OF THIS BOOK

This book is intended for tutors and/or parents who provide math instruction for public, private, or homeschooled students. It contains 60 lesson plans which correspond directly to the first two books of the series.

The book assumes that you as the instructor have a strong familiarity and comfort level with the material, and that you have basic skills in presentation. You don't necessarily have to be a professional teacher with years of experience, but keep in mind that your students will pick up on any doubt or insecurity on your part. It is also important that you have a good understanding of how all of the topics relate to one another. This is discussed more in the next chapter.

INTRODUCTION

It is important to understand that for the sake of conciseness, these lesson plans are very much abridged. They are intended as an overall framework and outline that the instructor can follow, or as a source of ideas. The lessons include tips on which concepts students tend to find easy or difficult, along with suggestions on which ones require reemphasis.

Like the rest of the series, this book is not intended to be used with students who are pursuing math-related degrees, or who want to explore the richness of the subject. The lessons were designed from the realistic standpoint that most students are just trying desperately to fulfill their math goals and requirements, and have very little time to do so.

HOW TO USE THIS BOOK

This book directly corresponds to the chapters of the first two books. Refer to those books in tandem with this one, using either purchased hardcopies or the free online versions.

Chapter One is entitled, "How to Be an Effective Math Instructor," and should not be skipped. In that chapter I have tried to incorporate all of the advice that I have to offer on the matter.

Chapters Two through Eleven contain 60 lessons which are directly aligned with the corresponding chapters in the main book and workbook. Start at Lesson 1 regardless of what level you believe the student is at, or what level s/he is according to government tests, or what level s/he says s/he is.

It is important to spend the appropriate amount of time on each lesson—no more and no less. This amount of time will vary for each lesson, which means that you as the instructor will have to be flexible and adaptive. If a student has truly internalized the material in a lesson, it may only be necessary to spend a few minutes on it, just for the sake of assessment and review. If a lesson involves totally new material, or material that the student finds confusing, it may require one or more hours, even in the context of private tutoring. Review previous lessons as needed, especially ones that contain prerequisite knowledge for the current lesson.

Chapter Twelve offers some advice on how you can provide meditation instruction to those students who are interested and able to take you up on your offer. While very few will, assuming you even want to offer it, the ones that do will have a distinct advantage in their math studies and in life in general.

THE FORMAT AND METHODOLOGY OF THE LESSONS

This book takes a standard, traditional approach to teaching math. Each lesson includes an aim or topic list, connections to earlier and/or later material, questions or problems to serve as a warm-up and motivation, demonstrative examples, points to elicit, and practice exercises which in most cases reference the companion workbook. Some of the lessons also include ideas for embellishment and/or points to reemphasize.

These lessons may be "dry" in compared to curricula that utilize non-traditional methods for teaching math. As I explain in the first book, most educators are very opposed to such special methods. It is important to understand that when a student takes a standardized exam, s/he is expected to solve problems quickly, and on his/her own. While this book does make reference to manipulatives and techniques of instructional merit, the student must not be dependent upon them.

INTRODUCTION

THE BOOK'S POSITION ON CALCULATOR USE

As described in the first book, this book takes a modern and practical view on calculator use. All but the very youngest students are typically permitted to use a calculator in class, for homework, and on exams. The trend in math education is for students to learn and apply concepts, and creatively make connections between them—it is not to see if students are able to do multi-step arithmetic by hand.

While working through the lessons, it is suggested that you adopt this practical view on calculator use, except where otherwise noted. For example, the student must learn to be capable of doing basic arithmetic computations by hand since doing them on a calculator is error-prone and time-consuming. The student should not use a calculator for tasks such as basic operations with fractions even if his/her calculator has such functionality. Again, it is the concepts and procedures which are important—not the actual "number crunching." To use a calculator for such tasks will only hinder the student.

HOW TO GET HELP WITH ANY OF THE LESSONS

I provide full support for this and the other books in the series via my comprehensive website at **www.MathWithLarry.com**. Feel free to e-mail me with any questions or comments that you have. I am happy to provide guidance on how to work with a student who is struggling with a particular topic. I can also provide additional practice exercises or demonstrative examples for any topic upon request, and of course I can answer any questions that you have about the material.

CHAPTER ONE

How to Be An Effective Math Instructor

THE TARGET AUDIENCE FOR THIS CHAPTER

This chapter is targeted at private, independent tutors, but the suggestions that are offered can be adapted for use by parents or homeschool instructors. To some extent, the suggestions can also be implemented by school teachers and by tutors who work for a commercial tutoring service. However, this chapter makes the assumption that the instructor has a fair amount of leeway and flexibility when working with students.

A DIFFERENT PERSPECTIVE ON MATH EDUCATION

This entire chapter is written from the standpoint that math education in America, for the most part, is implemented completely wrong. There may be valid reasons for why this is the case, and there may be political and economic reasons why changes to the system are slow in coming or non-existent, but it unfortunately is still the case.

In this chapter I share what I have learned from my experience with teaching and private tutoring. I have repeatedly seen first-hand what is effective for students, what is ineffective, and what appears on the surface as though it is effective, but in reality is not. As you read this chapter, understand that many of the suggested methods, while effective, will not be well-received by your students and/or their families.

ELICIT AND EXPLORE, DON'T LECTURE

The most essential technique of being an effective math instructor is to conduct your lessons by questioning and eliciting, rather than lecturing. Even at the college level it is extremely difficult to sit through a lecture and absorb information. Obviously this is even more the case for younger students and/or students who are not comfortable with or dislike math.

Throughout the book's lessons, I frequently use the word "elicit." What this means is to do your best to coax the given concept or information out of the student. The specific technique will vary for each situation, but it always involves forcing the student to think. This may involve recalling previous knowledge, or creatively applying concepts in new ways, or using logistical deduction, or even taking an educated guess. The point is not even whether the student answers your questions right or wrong. It is just a means of making the lesson student-centric as opposed to teacher-centric.

TECHNIQUES FOR ELICITING INFORMATION

There is no set formula for eliciting information. What is effective, though, is to constantly ask questions such as, "What do you think we should do next?," or "Why did I put the 3 over there?," or "What would we have done differently if...?," etc. Out of context such questions may come across as neurotic, but it is truly possible to conduct a lesson in this way without it being so. The worst that will happen is that the student will be frustrated by the fact that you are forcing him/her to think, as opposed to being permitted the luxury of just sitting there staring at you while nodding his/her head.

VISUAL LEARNERS: SHOW, DON'T TELL

It is very common for parents to inform their child's tutor that their child is a visual learner, and request that his/her tutoring be tailored to match this unique style of learning. The reality is that virtually everyone is a visual learner. As human beings, we learn by seeing. We learn even better by doing. Only a very small percentage of students, regardless of age, would say that they prefer to learn by way of a lecture as opposed to an interactive lesson full of various visual aids.

With that in mind, you should certainly make your lessons as interactive as you possibly can. Use drawings, diagrams, and manipulatives where appropriate, and keep any monologue-style instruction to a bare minimum. If at all possible, use a dry erase board and multicolored markers rather than pen and paper which forces the student to tilt his/her head.

Despite all of that advice, there is something you must make abundantly clear to students and their parents (if applicable). Not every single math topic lends itself to a special visual presentation. For many topics, the best that can be done from a visual standpoint is demonstrating steps on paper or a dry erase board. Some parents will insist that you teach a topic such as factoring a polynomial by way of manipulatives and/or elaborate drawings or graphics. It simply doesn't work that way, and it is a huge disservice to students as well as instructors that some textbook authors attempt to make it work by all sorts of contrived means.

Revisiting the example of factoring a polynomial, certainly you should illustrate the traditional "FOIL" arrows, and demonstrate how to list the various factor pairs which will be tested, and write all of the steps involved. However, that is where the visual aspect of the lesson ends. The use of manipulative tiles and similar only serve to ultimately hold the student back.

I have repeatedly seen students become very confused with such methods, even though on the surface they may seem as though they are beneficial. Even if a student does learn how to solve such problems using these special methods, s/he will be completely sunk on exams when the use of manipulatives are not permitted, and problems must be solved very quickly.

FLEXIBILITY IN LESSON CONTENT AND TIMING

Time-management and flexibility are huge components of effective tutoring. Your goal should be to make every minute of the tutoring session be as useful as possible. This involves having much more lesson material planned and prepared than you may get to. It also means understanding that even if you prepare very little material, you may only be able to cover a small amount of it. Your tutoring needs to be totally about the student, and not about any logistical concerns on your part.

Not every lesson will take the same amount of time, and the pacing within a lesson will vary from lesson to lesson. Depending on circumstances you may need to spend an entire lesson reviewing previous lessons, even though on the surface this may come across as "having done nothing" with the student. The goal must always be to learn math. That is accomplished by helping the student master the material one concept at a time, including all prerequisites for each concept.

WHY STUDENTS STRUGGLE WITH MATH

The first book devotes a fair amount of space to addressing the issue of why students struggle with math, but some points are worth reiterating. Students struggle with math because they move ahead to more advanced topics before they are truly ready to do so. In a school environment there are logistical and other reasons why this is the case, but it does not have to be so in the context of private tutoring or homeschooling.

It is very important that you not move past a topic until it is fully mastered. The attitude of, "The topic will come up again later" almost always results in the student declaring that "math is hard." Not only must you stay on a topic until it is fully internalized, you may also need to review and/or re-teach previous topics upon which the current topic is based. It is just that simple, and there is no true way around it.

TUTORING FOR THE SAKE OF GRADES

It is very common for students or their parents to seek out tutoring for the sake of grades, and/or averting the need for summer school. Furthermore, it is common for these requests to be made at "the last minute." In the first book I discuss how concerns about exams and grades are actually a huge part of the reason why students struggle with math. No math is actually learned when a teacher and/or tutor is doing nothing but "teaching to the test" in a last-minute, panicked frenzy.
As a tutor, you will need to make a decision about whether you will take on "last minute" clients, or if you will limit yourself to only long-term clients who want to learn math so they can end the pattern of constantly failing, or just barely passing only to be even more confused in the next grade. As I

say repeatedly and as I will say again, the only effective tutoring is that which is devoted to actually teaching math.

ASSESSING A STUDENT'S KNOWLEDGE

Many parents request that an assessment test be given to their child. There are significant pros and cons to doing this. One advantage is that by administering an assessment test, you have tangible proof of the student's level of math knowledge. This will give you some "ammunition" when you insist on reviewing prerequisite material which is essential for the material for which the student is requesting help.

A big disadvantage to administering an assessment test is that it may embarrass the student, and can be a "wake-up call" which the student and his/her parents are not prepared to receive. Many parents are convinced that their high school child is only "slightly behind," and are mortified to discover that their s/he cannot do simple fourth grade math tasks unless given extensive hints and help.

Another disadvantage is that if the student or his/her parents are insisting that you devote all of your tutoring time to homework help, and/or preparation for upcoming exams, then the result of the assessment test can easily become moot. There will not be any time to address the educational gaps which the assessment test uncovered.

FORCING THE STUDENT TO THINK

As previously described, if you want to be an effective tutor, it is absolutely essential that you force your students to constantly think. This may not make you popular with your students, and it may result in lost business. The choice is yours. What most students and their families don't realize is that math exams are really hardly about math. There aren't that many different concepts to learn, and the ones that must be learned are really not that difficult.

Modern exams are designed to test the extent to which the student is able to think independently and creatively, and apply concepts appropriately. The student will never develop those skills in a setting in which s/he is constantly told what step is next, or in which hints are provided to the extent that the s/he didn't have to do any thinking at all.

WORKING WITH LEARNING DISABLED STUDENTS

This important topic is addressed in the first book, but some points are worth reemphasizing in this book. If you follow all of the tutoring suggestions that are being made in this book, the entire concept of a learning disability will become a moot point. The overall tutoring process is the same for all students. You must spend as much time on a topic as necessary, and you must review previous topics to the extent that is necessary.

All off this will vary from student to student, but the guideline is still the same. This is also true for the way in which you provide instruction—it must be tailored for each student, and you must be flexible in offering alternative approaches to a topic. In plain English, if the student isn't grasping the concept in one way, you must try another. This is true for all students.

It is very important to understand that students can pick up on whatever you are thinking, and whatever you believe. If you have the attitude of, "Here goes another lesson with my

ADHD student," you can expect to observe a response that is characteristic of a student with ADHD. The same is true of any disability. We tend to live up to how others label us.

A huge problem in our culture in general is that we insist on applying labels to everything. You will be a highly effective tutor if you can get to the point where you just say, "My next student is Johnny. I'm going to help Johnny with his math by focusing on Johnny, and by adjusting my instruction as needed so that Johnny grasps the concepts." Notice there is no mention of any disability in that sentence. It's just "Johnny."

Remember, every single one of us has our disabilities and our special needs. Some are named, and some are not. Some have acronyms, and some do not. Just focus on each student as an individual. It's not "Johnny-The-ADHD-Kid," it's "Johnny."

"TEACHER, WHY DO I HAVE TO KNOW THIS?"

This question is very likely to come up at some point. My method of tutoring is to always be fully honest with students. For older students, the best and most honest answer to this question is that most of the math that s/he will learn in school does not have any direct application in everyday life. What does apply, though, is being able to solve problems by thinking analytically, independently, and creatively. That is what most employers and colleges are most concerned about, with the exception of careers and fields of study which require very specialized knowledge. Most students are willing and able to accept that math is the best subject for developing these skills.

For younger students, the best answer is that the student is too young to know for certain what s/he wants to do as a career, and what s/he wants to study in college. Remind the student that s/he is learning basic math, and that all students need to have a well-rounded education complete with a basic knowledge of all subjects. If a somewhat older student is absolutely convinced that s/he will never study or work with math, then try to adapt the first suggested response.

HOMEWORK HELP VERSUS FILLING IN GAPS

Aside from requests for exam preparation, the second most common request from parents is to help their children with their homework. The parent's mindset is typically, "Well, at least for one out of the five weekdays I'll get a break from helping which is good because I don't understand any of this crazy new math anyway."

As described, you will need to make a choice about what type of tutoring service you wish to offer. Helping with homework typically ends up taking an entire tutoring session. Even if the assignment is completed sooner, students often request help with previous assignments which they can resubmit for extra credit, and similar. Of course this does not leave any time left over to go back and review prerequisite knowledge which the student is lacking, and even if there was time, such a suggestion will likely be met with, "But I already did that like maybe four, no, wait, five years ago."

As with the matter of exam preparation, the matter comes down to how concerned you are about being popular, and how concerned you are about losing business. All I can say is that no math is learned by focusing on topics for which the student is not prepared. It may look as though you are helping the student to complete his/her homework assignments, and everyone may be all smiles at the end of each session, but nothing was truly learned as will be later evident.

ENSURING THAT THE STUDENT KNOWS WHAT TO EXPECT OUT OF TUTORING

In what may first appear as a change of mindset, the chapter concludes with the suggestion to always maintain a positive and optimistic atmosphere during tutoring sessions. The only way to accomplish this is by not tutoring in a manner which you know will be ineffective, and which you know in your heart is the proverbial "smoke and mirrors." You can also accomplish this by being fully realistic, and by not making false promises to the student or his/her parents.

For example, if a student comes to you for help preparing for the SAT, but barely knows what a fraction is, you won't accomplish anything in just a few last minute sessions. You won't even accomplish all that much in 10 sessions, if there is even time for that before the exam. There is just too much catching up to do, and the student's goals are unrealistic for his/her financial and time constraints. Only take on a student if you are certain that you can make observable progress with him/her. Make very clear that you are going to force him/her to constantly think, and that s/he does not expect to sit back and relax while you perform a one-man/woman monologue.

CHAPTER TWO

Lesson Plans for "The Foundation of Math: Basic Skills in Arithmetic"

Lessons 1 to 6

Topics Covered in This Chapter:

Basic concepts of addition, subtraction, multiplication, and division; Place value to thousands; Addition with carrying; Subtraction with borrowing; Multiplication of two-digits by one-digit

MATH MADE A BIT EASIER LESSON PLANS:
A GUIDE FOR TUTORS, PARENTS, AND HOMESCHOOLERS

LESSON 1

AIM/TOPIC(S): Basic concepts of addition

MAIN CONCEPT(S): Base-10 system; Place value to 1000; Commutativity; Addition representing combining; Addition facts involving one-digit numbers; Regrouping to facilitate mental math

PREREQUISITE KNOWLEDGE: The student must have at least a general familiarity with how numbers work—about a 2nd grade math knowledge or equivalent experience.

WARM-UP: Ensure that the student is comfortable counting by ones starting anywhere between 1 and 1000. Some students get confused when counting past numbers ending in 9, or that lead to the use of a new place. Many students spontaneously start counting by tens after reaching a multiple of 10 or 100.

MOTIVATION: Obviously addition is at the foundation of all math which will follow. Try to kindle interest by eliciting when the student uses or might use addition in everyday life, of course phrasing your question at an age-appropriate level.

POINTS TO ELICIT: Addition means combining. The answer is called the sum. We can combine two things in any order we want (define this as the commutative property). Multiplication is also commutative, but subtraction and division are not (this is covered in later lessons). Adding 0 to a value has no effect. Elicit the concept of our base-10 math system by discussing what happens when we count past 9, 99, and 999 (formalized later). Many students take for granted that we only have ten symbols to work with. Stress the concept of our place value system through the thousands place. Many students believe that it is OK to squeeze two digits into one place, which of course leads to confusion in later work.

DEMONSTRATIVE EXAMPLES:
Draw the positive side of a number line. Show how we can add by starting at a given number, and moving to the right the number of hops equal to the value to add. This is cumbersome and error prone, as is finger-counting. Demonstrate the practice and memorization of basic addition facts using homemade or store-bought flashcards. Elicit that 9+7 is the same as 10+6 whose sum is easier to compute mentally. Reinforce that we're just combining groups, so it's OK to move an object from one group into the other.

PRACTICE EXERCISES: Use flashcards to learn basic addition facts involving one-digit numbers. Emphasize that it is not practical to use a calculator for such tasks. Stress that once x + y is memorized, so too is y + x. Stress the concept of converting 9+7 to 10+6, or 8+5 to 10+3, etc.

IDEAS FOR EMBELLISHMENT: Depending on age level, practice using any type of manipulatives which the student will find interesting. Discuss ways (e.g., keywords/keyphrases) in which addition is implied in a word problem.

POINTS TO REEMPHASIZE: Many students are not fully convinced that addition is commutative. When adding 2 + 19, they will sooner start at 2 and count up 19 than vice-versa. Try to dispel this by any of the above means.

INSTRUCTOR'S NOTES:

MATH MADE A BIT EASIER LESSON PLANS:
A GUIDE FOR TUTORS, PARENTS, AND HOMESCHOOLERS

LESSON 2

AIM/TOPIC(S): Basic concepts of subtraction

MAIN CONCEPT(S): Subtraction representing taking away; Non-commutativity; 3 − 5 is not "unsolvable"; Facts involving one-digit numbers; Addition and subtraction fact families

CONNECTIONS TO EARLIER MATERIAL: The student must have fully mastered the previous lesson and its prerequisites.

WARM-UP: Quiz the student on basic addition facts involving one digit numbers. Flashcards are ideal for this purpose. The student should not show signs of hesitation or doubt, nor guess at each answer. Stress that on standardized exams and in the "real world," there is no partial credit for being "close."

MOTIVATION: Obviously subtraction is among the four basic operations in math. Try to kindle interest by asking the student when s/he uses or might use subtraction in everyday life, of course phrasing your question at an age-appropriate level.

POINTS TO ELICIT: Subtracting means taking away. The answer is called the difference. Subtracting 0 from a value has no effect. Subtraction is not commutative under any circumstances—stress this constantly. 3–5 cannot just be rewritten as 5–3, but that does not imply that the former is unsolvable. This is covered later. Addition and subtraction are inverse operations—they "undo" each other.

DEMONSTRATIVE EXAMPLES: Draw a positive number line, and show how we can subtract by starting at a given number, and moving to the left the number of hops equal to the value to subtract. This is cumbersome and error-prone, as is finger-counting. Demonstrate the practice and memorization of basic subtraction facts using homemade or store-bought flashcards. Elicit that we cannot regroup 9–7 as 10–6 since subtraction is not about combining.

Demonstrate (using counters if age-appropriate) that all of the following are in the same fact family: $2 + 5, 5 + 2, 7 - 5, 7 - 2$. Rather than compute something like $99 - 97$ by "taking away," it is much easier to mentally solve $97 + ? = 99$.

PRACTICE EXERCISES: Practice counting backwards by ones starting at different values through 1000. Use flashcards to learn basic subtraction facts involving one-digit numbers which do not lead to negative answers. Emphasize that it is not practical to use a calculator for such tasks.

IDEAS FOR EMBELLISHMENT: Depending on age level, practice using manipulatives which the student will find interesting. Elicit the various ways (e.g., keywords/keyphrases) in which subtraction can be implied in a word problem.

POINTS TO REEMPHASIZE: As mentioned, many students don't accept that 3–5 is neither "unsolvable" nor +2. Note that when performing two-digit subtraction, if we have such a scenario in the units place, the teacher will often say "we can't do it," and lead into the concept of borrowing. That concept will be covered in a later lesson. For now, have the student perform the problem on a number line by starting at 3 and counting 5 hops to the left. The matter is a "tough sell" for many students, but be certain to dispel the misconception.

INSTRUCTOR'S NOTES:

MATH MADE A BIT EASIER LESSON PLANS:
A GUIDE FOR TUTORS, PARENTS, AND HOMESCHOOLERS

LESSON 3

AIM/TOPIC(S): Basic concepts of multiplication

MAIN CONCEPT(S): Multiplication is commutative, and is a shortcut for repeated addition; Multiplication representing multiple groups; Facts through 12 × 12; Positive multiples

CONNECTIONS TO EARLIER/LATER MATERIAL:
The student must have fully mastered the previous lessons and their prerequisites. Being able to quickly and accurately list multiples is essential for later topics including fractions.

WARM-UP: Quiz the student on basic addition and subtraction facts. The student should not show signs of hesitation, doubt, or guessing. Create 7 groups of 8 items using any age-appropriate means, and ask the student to count them all. Use this to lead into the motivation described below.

MOTIVATION: Many students don't understand how multiplication is related to addition. Use the warm-up exercise to convince the student that we don't want to count all 56 objects one by one, nor do we even want to count by eights. By memorizing the multiplication table, we'll never have to "reinvent the wheel." Try to kindle interest by eliciting when the student uses or might use multiplication in everyday life.

POINTS TO ELICIT: Multiplication is just repeated addition. We repeatedly add a given number to itself a given number of times, and we'll never have to again if we memorize the result. The answer is called the product. Any value times 0 equals 0 since we have 0 groups. Any value times 1 is itself since it means one group of that value. Multiplication is commutative.

LESSON PLANS FOR "THE FOUNDATION OF MATH: BASIC SKILLS IN ARITHMETIC"

DEMONSTRATIVE EXAMPLES: Have the student draw a blank 13×13 multiplication table and label the row and column headers, along with an × sign in the upper left. Begin filling it in by computing one cell at a time. Stress that we're really just counting by a given number, e.g., 8, 16, 24, etc. Counters may be helpful for this. Emphasize the pattern of rows where applicable, especially for 2, 5, 10, and 11. If you sum the digits of any multiple of 9, it will add up to 9 or a multiple of 9. Elicit the symmetry of the rows and columns due to commutativity.

PRACTICE EXERCISES: Use flashcards to learn basic multiplication facts through 12×12. Emphasize that the multiplication table must be memorized. Practice listing multiples of numbers between 1 and 12, and quiz the student on whether a given number is a multiple of a particular number. Emphasize that this is important for basic problems involving fractions and later work with algebra.

IDEAS FOR EMBELLISHMENT: If age-appropriate, let the student work with counters to observe commutativity. Also work with quantities which have many different factor pairs such as 24, and see how many different ways the counters can be grouped. Elicit ways (e.g., keywords / keyphrases) in which multiplication can be implied in a word problem.

POINTS TO REEMPHASIZE: As mentioned, many students cannot accept that multiplication is just a shortcut for repeated addition. Do not allow the student to view it as some exotic operation. Stress that the practice of listing multiples of a number is not just busy work, but is used constantly in math.

INSTRUCTOR'S NOTES:

LESSON 4

AIM/TOPIC(S): Basic concepts of division

MAIN CONCEPT(S): Non-commutativity; Division as a shortcut for repeated subtraction, but better to think of as the inverse of multiplication, or just splitting into equally-sized groups; Fact families for multiplication and division; Representations of division; Remainders using "R" notation.

CONNECTIONS TO EARLIER / LATER MATERIAL:
The student must have mastered the previous lessons and their prerequisites. Obviously division is key for later work.

WARM-UP: Quiz the student on basic addition, subtraction, and multiplication facts. Ask the student to list the first 12 multiples of selected numbers between 1 and 12. The student should not show signs of hesitation, doubt, or guessing. Pick a value that factors in different ways such as 48, and use an age-appropriate means for the student to experiment with different ways that s/he can form equally-sized groups.

MOTIVATION: Stress that there is often the need to divide a quantity into groups of equal size, perhaps with some items remaining. Try to kindle interest by asking when the student uses or might use division in everyday life. Emphasize that division is not an exotic operation that someone invented, nor is it a new operation to learn. It is nothing more than the inverse of multiplication.

POINTS TO ELICIT: The student learned that subtraction is the inverse of addition since it "undoes" it. Division and multiplication parallel this. The solution to a division problem

LESSON PLANS FOR "THE FOUNDATION OF MATH: BASIC SKILLS IN ARITHMETIC"

$b\overline{)a}^{c}$ is the quotient. The quantity being divided is the dividend, and that which is doing the dividing is the divisor. Division is based on equally-sized groups. Anything left over is a remainder. The diagram at left represents $a \div b = c$. Many students confuse what goes where.

DEMONSTRATIVE EXAMPLES:
Use counters or similar means to show that the following are a fact family: 8×3, 3×8, 24÷3, 24÷8. Show how we can use the multiplication table "in reverse" to solve division problems by finding the row of the divisor, moving across to the cell with the dividend, then moving up to the corresponding column header to get the quotient. Explain why n ÷ n = 1 and n ÷ 1 = n.

Demonstrate the use of the "R" notation in 23 ÷ 7 = 3 R 2. Division is not commutative—dividing 24 candies among 3 people is not the same as dividing 3 candies among 24. If the student asks, explain that this is covered in the unit on fractions, but it is not the case that 3 ÷ 24 is "unsolvable."

0 ÷ n and n ÷ 0 are covered in a later lesson, but can be addressed if the student asks. 0 ÷ n = 0 because we're distributing 0 items. For now it may be best to just say that n ÷ 0 is "undefined" (show error on calculator), and will be explained later.

PRACTICE EXERCISES: Use flashcards to learn basic division facts that can be answered via the multiplication table.

IDEAS FOR EMBELLISHMENT: Elicit ways (e.g., keywords/keyphrases) that division can be implied in a word problem.

INSTRUCTOR'S NOTES:

MATH MADE A BIT EASIER LESSON PLANS:
A GUIDE FOR TUTORS, PARENTS, AND HOMESCHOOLERS

LESSON 5

AIM/TOPIC(S): Place value to thousands; Addition (carrying)

CONNECTIONS TO EARLIER MATERIAL: The student must have mastered the previous lessons and their prerequisites.

WARM-UP: Quiz the student on the basic four operation facts using flashcards or similar. Ask the student to list the first 12 multiples of selected numbers between 1 and 12. The student should not show signs of hesitation, doubt, or guessing.

MOTIVATION: Even if calculators are permitted in class and on exams, it is essential to be able to do some basic calculations by hand. With this we'll review place value concepts.

POINTS TO ELICIT: Have the student count by ones starting at 7, 97, and 997. Elicit the concept of our base-10 system where each place is worth 10 times as much as the place value on its right. We only use 10 different symbols (digits) whose values are dependent upon which column (place) they are in.

DEMONSTRATIVE EXAMPLES:
Ask the student why 847 and 784 are not equal, even though they have the same digits. Compare the place value chart to a cash register with compartments for bills of $1, $10, $100, and $1000. Students will ask about $5 and $20 bills, but explain that our math system only uses values of 1, 10, 100, 1000, etc. Explain that our special cash register cannot have more than 9 bills in a compartment. Once we have more than that, we must change them in and "trade up" to the next higher denomination. Use real/play money or other manipulatives as age-

appropriate. Explain that we use a 0 as a placeholder to represent no bills of a given denomination.

Demonstrate the addition of two-digit by two-digit numbers which leads to carrying (e.g., 37 + 45). Many students don't fully accept the "one digit per place" rule, and answer the above problem as 712. Use the cash register analogy of only being allowed 9 bills in a compartment. We must trade in 10 of our 12 singles for a $10 bill. Show how we carry into the next higher column. Use manipulatives as needed.

Stress that we add the columns by working right to left since we'll carry into higher place values on the left. Many students balk at working right to left, and even at adding by columns. Stress that we can rewrite 10 ones as 1 ten, but we can't rewrite 10 tens as 1 one. When we get to the tens place, stress that we are really adding 30 + 40, plus the 1 ten which we carried.

Demonstrate a problem such as 79 + 56 which will lead to an answer in the hundreds place. Show that we end up with 13 tens which is really 130, or 3 tens and 1 hundred. Have the student count 13 groups of 10. We would carry the 1 if we had anything to add in the hundreds place, but since we don't, we can just write it in the answer.

PRACTICE EXERCISES: Practice adding a variety of two-digit by two-digit numbers, including some which don't require carrying. Some students start carrying "by rote," not realizing its significance and when it is not applicable. Insist that the student explain every step along the way, and account for what each digit actually represents.

INSTRUCTOR'S NOTES:

MATH MADE A BIT EASIER LESSON PLANS:
A GUIDE FOR TUTORS, PARENTS, AND HOMESCHOOLERS

LESSON 6

AIM/TOPIC(S): Subtraction with borrowing; Multiplication of two-digits by one-digit

CONNECTIONS TO EARLIER MATERIAL: The student must have mastered the previous lessons and their prerequisites.

WARM-UP: Quiz the student on the basic four operation facts using flashcards or similar. Ask the student to list the first 12 multiples of selected numbers between 1 and 12. The student should not show signs of hesitation, doubt, or guessing. Quiz the student on two-digit plus two-digit addition.

MOTIVATION: Even if calculators are permitted in class and on exams, it is essential to be able to do some basic calculations by hand. With this we'll review concepts of place value.

DEMONSTRATIVE EXAMPLES / POINTS TO ELICIT:
Demonstrate a problem like 87 – 24. As with addition, we subtract the columns from right to left. In this case there are no complications. Now change the problem to 87 – 29 and ask what the point of concern is. Many students have no qualms about swapping the 7 and 9 to get 9 – 7 = 2. Obviously this has to be completely nipped in the bud immediately. Use real/play money or counters to demonstrate how we borrow 1 ten from the tens place, and return it in the form of 10 ones, effectively turning the 7 into 17. Many students are not convinced that all of this is "legit," so let them keep working with manipulatives.

Demonstrate a multiplication problem like 34 × 2. Explain how we essentially distribute the 2 over both the 3 and the 4. We start by multiplying 2 times 4 to get 8 which can go in the ones place of the answer. Then we multiply 2 times 3, keeping in mind that the 3 represents 30. We get 6 which really means 60 or 6 tens which can go in the tens place of the answer.

Now demonstrate a problem like 68 × 7. Multiply the unit place values as above to get 56. Emphasize that we can never put more than one digit in a place. We can write the 6 ones in the ones place of the answer, but we must carry the 5 into the tens place. Stress that this reminds us we still must account for 5 tens or 50 which will get added in later. Some students get confused because that step involves adding.

Continuing, we multiply 7 times 6 to get 42, which is really 420 or 42 tens (remembering that the 6 meant 60). We must add in the 5 carried tens which were left unaccounted for, giving us 47 tens which can go in the answer to the left of the 6. Stress that the 47 tens are really 7 tens and 4 hundreds which is why the 4 ended up in the hundreds place.

PRACTICE EXERCISES: Practice subtracting a variety of two-digit by two-digit numbers, including some that don't require borrowing. Some students start borrowing "by rote," not realizing its significance and when it is not necessary. Insist that the student explain every step along the way, and account for what each digit actually represents.

Following the same guidelines, practice multiplying a variety of two-digit by one-digit numbers, including some which don't require carrying.

INSTRUCTOR'S NOTES:

CHAPTER THREE

Lesson Plans for "Basic Math Topics and Operations"

Lessons 7 to 14

Topics Covered in This Chapter:

Integers; Even/Odd numbers; Greater/Less than; Exponents; Perfect squares/cubes; Powers of 0, 1; Square roots; Prime and composite numbers; Factors; Order of operations (PEMDAS); Distributive property; Place value to billions; Reading/writing numbers with words; Rounding whole numbers

MATH MADE A BIT EASIER LESSON PLANS:
A GUIDE FOR TUTORS, PARENTS, AND HOMESCHOOLERS

LESSON 7

AIM/TOPIC(S): Integer definition; Determining if a computation results in an even or odd value; Greater/Less Than

CONNECTIONS TO LATER MATERIAL: The term integer comes up constantly in math, and but will not be defined when it does. Its definition must be memorized.

WARM-UP / MOTIVATION: Give the student an assortment of even and odd quantities (consider using counters), and ask him/her to divide each by 2. Ask for his/her findings. Ask the student why 703 is greater than 73, even though "0 means nothing." Ask why 987 is less than 1234 even though the latter is comprised of smaller digits. Ask the student for examples of what s/he considers to be whole numbers.

DEMONSTRATIVE EXAMPLES / POINTS TO ELICIT:
We define "integer" (the "g" is pronounced like "j") as a whole number. It can be positive, negative, or 0. Have the student give some representative examples. If the student's math background permits, discuss why ½ and 0.5 are not integers.

We define an even number as a number that ends in 0, 2, 4, 6, or 8. An even number can be divided by 2 with no remainder. Revisit the related warm-up exercise. Numbers that are not even are called odd. An odd number ends in 1, 3, 5, 7, or 9. In each case it is only the rightmost digit that is of concern. Stress that 135,792 is an even number. Elicit that when any odd number is divided by 2, it will have a remainder of 1.

Present the ">" and "<" symbols for greater than and less than. The popular analogy of an alligator's mouth opening to the bigger quantity works well for all ages. Revisit the related examples from the warm-up to review the concept of place value. If the student seems hesitant, put a dollar sign to the left of the values in question, and ask which represents the amount of money s/he would prefer to be given. Most students are fine with this topic, and only run into trouble when attempting to compare fractions and decimals.

PRACTICE EXERCISES:
A popular question is of the form, "Will the result of computing Even × Odd × Even be even or odd?" The actual words "Even" and "Odd" will be spelled out. While there are rules for computations of this nature, they are hardly worth memorizing. Elicit that a very simple odd number is 1, and a very simple even number is 2. When faced with a problem like this, just use those values in place of the words "Odd" and "Even," and see if the answer is even or odd. Whatever answer you get is the answer to the problem.

Most students do fine with with these topics which will likely result in this lesson taking up less time than others. Consider using any extra time for review by quizzing the student on previous material.

POINTS TO REEMPHASIZE: The student must memorize the complete definition of integer. Stress that the term comes up constantly in the instructions to problems and in word problems. On formal exams, the proctor will not be permitted to define it for the student, and not understanding it can often mean getting an otherwise easy question wrong.

INSTRUCTOR'S NOTES:

MATH MADE A BIT EASIER LESSON PLANS:
A GUIDE FOR TUTORS, PARENTS, AND HOMESCHOOLERS

LESSON 8

AIM/TOPIC(S): Exponent terminology; Exponents as repeated multiplication; Perfect squares; Cubes; Power of 0 and 1

CONNECTIONS TO LATER MATERIAL: These skills and concepts are used constantly in later math, especially algebra. There is no point in progressing until they are mastered.

WARM-UP / MOTIVATION: Ask the student to compute 3×3×3×3×3×3×3×3 (eight copies), ideally without a calculator. Discuss how easy it is to make a mistake, or to lose count of where you are, or to miscount the number of copies involved. Now ask the student to compute 3+3+3+3+3+3+3+3 (eight copies). Use these results to lead into the lesson

DEMONSTRATIVE EXAMPLES / POINTS TO ELICIT:
Ask the student if there is any shortcut representation for the second warm-up problem. Remind the student that multiplication is a shortcut for repeated addition. Ensure the student recognizes that the first problem deals with repeated multiplication which admittedly doesn't have as many everyday applications as repeated addition. Elicit the student's ideas about why the answer to the first problem was so much larger than the second problem.

Explain that we have a shortcut notation and shortcut operation for repeated multiplication. Elicit the connection between the first warm-up exercise and 3^8. Define the 8 as the exponent or power, and define the 3 as the base. Ensure the student understands that the two are not interchangeable, and that the exponent must be written as a superscript.

LESSON PLANS FOR "BASIC MATH TOPICS AND OPERATIONS"

Ensure the student recognizes that 3^8 means to take the base and multiply it times itself for a total of 8 copies. It does not mean to add 3 to itself 8 times which would be 3 × 8. Be certain that the student does not think that 3^8 is a misprinted 38.

Explain that exponents of 2 and 3 are very common in math and in geometry, and are usually read as "squared" and "cubed," respectively. If the student is interested, briefly digress to the related geometry of squares and cubes, and if not, just reinforce that n^2 = n × n, and n^3 = n × n × n.

Elicit that n^1 = n, and ask the student to memorize that n^0 = 1 with the explanation saved for later work in algebra.

PRACTICE EXERCISES:
It is essential for later work that the student memorize the perfect squares from 1^2 to at least 12^2, ideally including the other common squares from the chart in the first book. This must be drilled regularly, with the student being able to square given integers, as well as indicate what integer must be squared to obtain a given perfect square. Ideally the student should memorize perfect cubes from 1^3 to 5^3.

If the student will be using a calculator for classwork and exams, ensure that s/he knows how to use the exponent function which varies depending on model. Demonstrate the use of the "^" notation which is commonly used when superscript notation is not available.

POINTS TO REEMPHASIZE: This entire lesson is extremely important for later math. Ensure that the student doesn't think of exponents as trivial "busy-work."

INSTRUCTOR'S NOTES:

LESSON 9

AIM/TOPIC(S): Square roots as the inverse of squaring

WARM-UP / MOTIVATION: Choose random perfect squares, and ask the student what value must be squared to obtain each one. Ensure that the student's answer does not include the word "squared" since the question itself implies squaring.

DEMONSTRATIVE EXAMPLES / POINTS TO ELICIT:
Illustrate the use of the square root symbol as in $\sqrt{64}$ ("The square root of 64.") The problem is asking us to state the number which must be squared in order to get 64. Elicit the inverse relationship between squaring and "square rooting." This is confusing for many students. It is common for students to answer 8^2. Explain that the question is just asking for the base. This sometimes leads to an unintended "sketch" in which the student will say, "But you said that 8^2 is 64." Explain that the "squared" part is built into the question itself (i.e., the "$\sqrt{}$" symbol), and an answer of 8^2 on a test will be marked wrong.

It is common for students to evaluate $\sqrt{64}$ as $\sqrt{8}$ since "we're doing a unit on square roots." Obviously dispel the misconception. Some students will evaluate as $\sqrt{64}$ as 4096 (squaring the 64). This is more likely for $\sqrt{4}$. Reinforce that squaring and "square rooting" are inverses and are not interchangeable.

PRACTICE EXERCISES: Quiz the student on evaluating the square roots of common perfect squares (i.e., the inverse of the last lesson). Build a table of the results and memorize.

INSTRUCTOR'S NOTES:

LESSON PLANS FOR "BASIC MATH TOPICS AND OPERATIONS"

LESSON 10

AIM/TOPIC(S): Prime and composite numbers; Factors

WARM-UP / MOTIVATION: Select an assortment of prime and composite numbers. For each number, ask the student to list all of the ways that the number can be obtained by multiplying two numbers together. Use this to lead into the lesson.

DEMONSTRATIVE EXAMPLES / POINTS TO ELICIT:
Define factor by example, noting that the factors of 36 are 1, 2, 3, 4, 6, 9, 18, and 36. Elicit the concept of factor pairs, noting that perfect squares have a factor which gets paired with itself.

Elicit that the prime numbers only had 1 and themselves as factors, and define such numbers as prime. Elicit that the composite numbers have at least one factor other than 1 and itself, and define such numbers as composite.

Ask the student if all even numbers are composite. Try to elicit that all even numbers will have at least three factors (1, 2, and themselves), with the exception of 2 which is prime. Ask the student if 1 is a prime or a composite number. S/he will likely answer prime. Use this to explain the "fine print" that a prime number must have two unique factors, 1 and itself. This makes 1 a special case (neither prime nor composite).

PRACTICE EXERCISES: Ask the student to list all the prime numbers through 100. Make a table of numbers 1 to 100, and discuss efficient ways of eliminating non-primes (e.g., multiples of 2, 3, 5, etc.).

INSTRUCTOR'S NOTES:

MATH MADE A BIT EASIER LESSON PLANS:
A GUIDE FOR TUTORS, PARENTS, AND HOMESCHOOLERS

LESSON 11

AIM/TOPIC(S): Order of operations (PEMDAS)

CONNECTIONS TO LATER MATERIAL: These skills and concepts are used constantly in later math, especially algebra. There is no point in progressing until they are mastered.

WARM-UP / MOTIVATION: Ask the student to evaluate the expression $2 + 3 \times 4$. Most will get 20. Tell the student that the answer is actually 14, and use this to lead into the lesson.

DEMONSTRATIVE EXAMPLES / POINTS TO ELICIT:
Explain that we are sometimes asked to evaluate an expression with many different operations. We need to have standard rules to follow which tell us the order in which to simplify it. Explain that such rules have been established, and they are identical all over the world. The rules are not just cultural, or something that the teacher or textbook invented. Many students ask why they just can't work left to right. Try to have him/her accept that it just doesn't work like that.

Present the acronym PEMDAS. Try to elicit what each letter stands for. Explain that we must always handle anything in parentheses before anything else, no matter how tempting. Present an example like $1 + (99 \times 2)$. Demonstrate working inner to outer in a problem such as $3 + [2 \times (6 - 1)]$. After parentheses, we handle exponents (and square roots).

The next level of precedence is multiplication and division. It is essential to stress that these operations share the same priority level, and are resolved left to right in the order that they appear. Present an expression like 10 ÷ 2 × 3. Many students say "M before D," and end up with 10 ÷ 6 for an answer of 1R4. Note that you may run into a bit of, "But teacher said!" Try to dispel the misconception, remembering that even the student's teacher may not fully understand it.

After all multiplication and division have been handled left to right, the next level of precedence is addition and subtraction. It is essential to stress that these operations share the same priority level, and are resolved left to right in the order that they appear. Present an expression like 10 − 4 + 5. Many students say "A before S," and end up with 10 − 9. Again, you may run into, "But teacher said!," which you must counter.

Present an expression such as 10 − 3 × 2. Stress that we must jump straight to the multiplication even though the subtraction occurs first. M and D always come before A and S.

Demonstrate how we should solve PEMDAS problems by working step-by-step and line-by-line, drawing arrows to help keep track of what each simplification step accomplished. Once students have accepted the above concepts, the biggest hurdle is getting students to not rush and combine steps.

PRACTICE EXERCISES / POINTS TO REEMPHASIZE:
The companion workbook offers some exercises that the student can work through, and of course you can make up your own. Stress that an answer obtained by ignoring or misapplying the PEMDAS rules is wrong. Show that basic calculators cannot be used for PEMDAS problems since they effectively compute left to right regardless of operation.

INSTRUCTOR'S NOTES:

LESSON 12

AIM/TOPIC(S): Distributive property of multiplication over addition and subtraction

NOTE TO INSTRUCTOR: Prior to algebra, this lesson is a hard sell. Just try to convey the general concept. Students don't see the purpose of the distributive property, nor are they convinced that at some point in the future it will be essential when we have unlike terms in parentheses.

WARM-UP / MOTIVATION: Ask the student to multiply 102 × 9 using the standard method. See if s/he can think of a way that the problem can be "split up" so that the answer can be computed mentally and quickly.

DEMONSTRATIVE EXAMPLES / POINTS TO ELICIT:
Revisiting the warm-up, have student compute 100 × 9 by any means. S/he will likely answer without hesitation. Ask him/her to compute 2 × 9 which should be easy. Elicit that there is no harm in rewriting 102 as (100 + 2), and that we can rewrite the original problem as 9 × (100 + 2). Granted we can begin by evaluating what is in parentheses which is what PEMDAS actually tells us to do, but we can also just distribute the 9 over the 100 plus the 2, giving us 900 + 18 which is 918.

Demonstrate distributing multiplication over subtraction such as 72 × 99 which becomes 72 × (100 − 1), leading to 7200 − 72.

PRACTICE EXERCISES: Practice problems which parallel these examples, with an emphasis on the addition ones.

INSTRUCTOR'S NOTES:

LESSON PLANS FOR "BASIC MATH TOPICS AND OPERATIONS"

LESSON 13

AIM/TOPIC(S): Place value to billions; Reading and writing large numbers with words

WARM-UP / MOTIVATION: Ask the student to name the first four whole number place values from right to left. Elicit that each place is 10 times the value of the place on its right. See if the student can continue the pattern through billions. If s/he can't, that will serve as the motivation. Many students are unmotivated to learn writing and recognizing large numbers in words. Stress that easy test questions can be gotten wrong because of failure to recognize a spelled-out number.

DEMONSTRATIVE EXAMPLES / POINTS TO ELICIT:
Many students are confused with the concept of one million. Elicit that it is 10×100,000 and 1000×1000. Elicit that one billion is 1000×1,000,000. Explain trillions if asked, and that zillions is fictitious. Show the use of the comma for ease of reading, with a comma placed after every three place values counting from the right (sometimes omitted for numbers in the thousands).

Demonstrate the proper format for writing numbers with words. The most important concept is that we do not use the word "and" for whole numbers because it will later be used to signify a decimal point. Show how we treat the three "units" places as a group, followed by the three "thousands" places, followed by the three "millions" places, etc.

PRACTICE EXERCISES: The companion workbook offers related exercises. Supplement with your own as needed.

INSTRUCTOR'S NOTES:

MATH MADE A BIT EASIER LESSON PLANS:
A GUIDE FOR TUTORS, PARENTS, AND HOMESCHOOLERS

LESSON 14

AIM/TOPIC(S): Rounding whole numbers to various places

WARM-UP / MOTIVATION: Quiz the student on the previous lesson on place value. Elicit the student's thoughts on what rounding is, when it might be used. Most students understand the general idea, and only need to learn or review the rules.

DEMONSTRATIVE EXAMPLES / POINTS TO ELICIT:
Let's round 78,625 to the nearest hundred. The first book outlined this procedure: Circle the place value to which we must round. Call it the target place. Underline the digit to the right of the target place. Call it the check digit. If the check digit is 4 or lower, round down by making the check digit and everything to its right 0. If the check digit is 5 or higher, round up by increasing the target place digit by 1, and making everything to its right 0. In this scenario, if the target place digit is 9, it becomes 0 (i.e., 10, with the 1 carried to the left).

PRACTICE EXERCISES: The companion workbook offers related exercises. Supplement as needed. Be sure to cover all place values between tens and billions, and include an assortment of rounding up and down. Include "tricky" problems such as rounding 199,502 to the nearest thousand.

POINTS TO REEMPHASIZE: Many students view rounding as just busy work, or as simple practice with place value. Stress the importance of using rounding to estimate answers before starting work on problems, as well as using it to verify that the obtained answer is reasonable.

INSTRUCTOR'S NOTES:

CHAPTER FOUR

Lesson Plans for "Working with Negative Numbers"

Lessons 15 to 20

Topics Covered in This Chapter:

Negative numbers; Signed number arithmetic (addition, subtraction, multiplication, division); Absolute value of constants and expressions; Squares and square roots of negative numbers

MATH MADE A BIT EASIER LESSON PLANS:
A GUIDE FOR TUTORS, PARENTS, AND HOMESCHOOLERS

LESSON 15

AIM/TOPIC(S): Introduction to negative numbers

CONNECTIONS TO LATER MATERIAL: Obviously negative numbers are used constantly in later math. They must be fully mastered now before entering the abstract world of algebra. Later lessons extend the basic four operations to negatives.

WARM-UP / MOTIVATION: Elicit the student's ideas on the concept of negative numbers. Responses may greatly vary including full understanding, minor misconceptions, and the notion that negative numbers are fictitious or theoretical. Use any response as the motivation for the lesson.

DEMONSTRATIVE EXAMPLES / POINTS TO ELICIT:
Draw a number line from about -5 to 5. Elicit its "mirror image" characteristic. Most students take well to thinking about negative numbers as debts, and positive numbers as assets, although they may not use those terms.

Elicit that we use the term "signed number" to mean a positive or negative number, but that positive signs are usually omitted. The most important goal of this lesson is to dispel any aversion that the student may have toward negative numbers.

PRACTICE EXERCISES: Practice simple addition and subtraction on the number line (i.e., counting hops left or right) to incorporate an assortment of positive and negative numbers and answers. It is essential to convince the student that 3 – 5 is not "unsolvable," nor do we convert to 5 – 3 to get 2.

INSTRUCTOR'S NOTES:

LESSON PLANS FOR "WORKING WITH NEGATIVE NUMBERS"

LESSON 16

AIM/TOPIC(S): Adding signed numbers

CONNECTIONS TO LATER MATERIAL: Mastery of this topic is absolutely essential for the later work including the study of algebra. This is one of the most important lessons in the book.

MOTIVATION: Ask the student what happens if s/he has $7 but owes someone $4. Then ask what happens if s/he owes someone $5 but only has $2. Use this to lead into the lesson.

DEMONSTRATIVE EXAMPLES / POINTS TO ELICIT:
After the last lesson the student should have a good feel for negative numbers. There are many ways to teach signed number addition, but an effective way is to have the student compare "what they have" with "what they owe." Most students grasp this intuitively. Try to elicit the models below which represent the four possible combinations that can occur.

An example like 3 + 5 can be modeled as, "You have $3 in one bank account and $5 in another," or any equivalent scenario. Just add up your assets (i.e., "what you have").

An example like (-2) + 7 can be modeled as, "You have $7 but owe someone $2 which you want to repay," or any equivalent scenario. Compute the difference between what you have and what you owe. Then determine if you owe more than you have, or vice-versa. If the former, the answer is negative since you will still be in debt after repaying all that you can.

Lesson continued on next page →

An example like 3 + (-5) can be modeled as, "You have $3 but you owe someone $5 which you want to pay back," or any equivalent scenario. Compute the difference between what you have and what you owe, and then determine if you owe more than you have, or vice-versa. In this case we can pay back the $3 that we have, but we still owe $2, represented as -2.

An example like (-4) + (-5) can be modeled as, "You owe one friend $4 and you owe another friend $5. In total you owe $9, represented as -9. We added up our debts, and are getting deeper into negative territory.

PRACTICE EXERCISES: The companion workbook has practice exercises on this topic. Supplement with your own, ensuring that all four combinations above are practiced. Remind the student that addition is commutative.

POINTS TO REEMPHASIZE: Reemphasize the entire lesson. Do not move past this lesson until the student can easily perform problems of this nature. The next lesson on subtraction refers directly back to this lesson, and of course the study of algebra requires mastery of these skills.

NOTE TO INSTRUCTOR: The student's textbook may teach this topic with a seemingly totally different method, but of course it is equivalent. Explore this with the student, and explain that many topics in math are of this nature. The important point is that any method for signed number addition ultimately comes down to negatives offsetting positives.

INSTRUCTOR'S NOTES:

LESSON PLANS FOR "WORKING WITH NEGATIVE NUMBERS"

LESSON 17

AIM/TOPIC(S): Subtracting signed numbers

CONNECTIONS TO EARLIER MATERIAL: The method in this book involves converting signed number subtraction problems into equivalent signed number addition problems. It is highly suggested that you use this method, but also explore any method which the student's textbook uses. Show that this method actually incorporates the popular "double negative" shortcut which converts 7 – (-2) into 7 + (+2).

WARM-UP / MOTIVATION: Ask the student to compute 7 – 4 by "moving left" on a number line. Then try computing (-3) – 2 on a number line. Then try 4 – 6. Then try 1 – (-2). This will either lead to confusion, or the student referencing the "double negative" shortcut. Then try (-3) – (-5). This will likely result confusion and/or guessing, which will lead into the lesson.

DEMONSTRATIVE EXAMPLES / POINTS TO ELICIT:
There are many ways to teach this topic. The first book outlined a simple four-step method which will always work.

1) Leave the first number alone.
2) Change the subtraction operation to addition.
3) Change the sign of the second number.
4) Compute the resulting equivalent addition problem.

PRACTICE EXERCISES: The companion workbook offers practice exercises, and of course supplement with your own. Don't move ahead until this topic is fully mastered.

INSTRUCTOR'S NOTES:

MATH MADE A BIT EASIER LESSON PLANS:
A GUIDE FOR TUTORS, PARENTS, AND HOMESCHOOLERS

LESSON 18

AIM/TOPIC(S): Multiplying and dividing signed numbers

CONNECTIONS TO LATER MATERIAL: This topic is essential for all later math. Don't move past it until it is mastered.

WARM-UP / MOTIVATION: Ask the student how much s/he would owe if s/he owed $4 to each of 7 friends. What sign would we used to represent this answer? Elicit that the problem can be solved using either addition or multiplication. Now ask the student to compute (-3) × (-5). This will likely lead to confusion or guessing which will lead into the lesson.

DEMONSTRATIVE EXAMPLES / POINTS TO ELICIT:
The following rules must be presented and memorized:

$$\text{Pos} \times \text{Pos} = \text{Pos} \qquad \text{Pos} \div \text{Pos} = \text{Pos}$$
$$\text{Pos} \times \text{Neg} = \text{Neg} \qquad \text{Pos} \div \text{Neg} = \text{Neg}$$
$$\text{Neg} \times \text{Pos} = \text{Neg} \qquad \text{Neg} \div \text{Pos} = \text{Neg}$$
$$\text{Neg} \times \text{Neg} = \text{Pos} \qquad \text{Neg} \div \text{Neg} = \text{Pos}$$

A positive times a negative can be thought of as a repeated debt. Real-world analogies for Neg × Neg come across as contrived, so the rule is best memorized. A helpful mnemonic device which students like is, "Matching is good (positive), and mismatched is bad (negative). Show how the rules for division are exactly the same as the ones for multiplication.

PRACTICE EXERCISES: The companion workbook offers practice exercises. Optionally supplement with your own. Don't move ahead until this topic is fully mastered.

INSTRUCTOR'S NOTES:

LESSON PLANS FOR "WORKING WITH NEGATIVE NUMBERS"

LESSON 19

AIM/TOPIC(S): Absolute value of constants and expressions

CONNECTIONS TO EARLIER/LATER MATERIAL: This topic serves as a good review of many previous topics. Most students find it extremely easy, and should be informed that it does play an important role in more advanced later math.

WARM-UP / MOTIVATION: Ask the student how far (how many hops) the number 7 is from 0 on the number line. Then ask how far -4 is from 0. Elicit the idea that distance is always positive regardless of the direction in which it is measured.

DEMONSTRATIVE EXAMPLES / POINTS TO ELICIT:
Explain that we have a simple operation in math which asks us for a value's distance from 0. As described, the answer will always be positive. Demonstrate the vertical bar notation as in $|7| = 7$, and $|-4| = 4$. Demonstrate some simple PEMDAS expressions with the vertical bars, explaining that we must first evaluate the expression, and then take the absolute value. An example is $|3 + (-4) \times 5| = 17$.

PRACTICE EXERCISES: The companion workbook offers practice exercises, and of course supplement with your own. Since this lesson is very short and simple, consider using any extra time to either move ahead, or spiral back to previous topics for review. By far the most important topics covered so far are the ones involving PEMDAS and signed number arithmetic. Return to these topics consistently, and be alert for guessing or hesitation on the part of the student.

INSTRUCTOR'S NOTES:

MATH MADE A BIT EASIER LESSON PLANS:
A GUIDE FOR TUTORS, PARENTS, AND HOMESCHOOLERS

LESSON 20

AIM/TOPIC(S): Squares and square roots of negative numbers

CONNECTIONS TO EARLIER MATERIAL: We've already worked with squares and square roots as applied to positive numbers. This lesson extends those topics to negatives.

WARM-UP / MOTIVATION: Ask the student to evaluate $\sqrt{16}$. Ideally the student will answer 4. Ask the student if there is another number which can be squared to get positive 16, and ensure him/her that there is. A common response is "8 × 2." Review the lessons on squares and square roots as needed.

Ask the student to evaluate $\sqrt{-16}$. We're looking for a number which when squared is -16. Responses will vary. Review the lesson on signed number multiplication as needed.

Ask the student to evaluate $(-7)^2$ and -7^2. Elicit his/her ideas on whether there is any difference. Use the warm-up exercises to lead into the lesson regardless of the student's responses.

DEMONSTRATIVE EXAMPLES / POINTS TO ELICIT:
Getting back to $\sqrt{16}$, elicit that a negative times a negative is positive. Elicit that if we square -4 we get +16 which implies that -4 is the square root of 16. Square root problems always have two solutions—a positive and matching negative version. Demonstrate the ± notation as in ±4 (plus or minus 4). Explain that "by default," we answer square root problems using only the positive version, known as the principal root.

Getting back to $\sqrt{-16}$, elicit that it is impossible to square a number and end up with a negative result. Explain that the

LESSON PLANS FOR "WORKING WITH NEGATIVE NUMBERS"

answer to such a problem is the word "Undefined." Many students have a hard time accepting this, so give the student extra time to convince him/herself.

Occasionally a student will have had exposure to more advanced math in which we have ways of working with negative square roots. Explain that unless the student goes on to study much higher level math, the answer is "Undefined."

Getting back to $(-7)^2$ and -7^2, explain that both of these are actually PEMDAS problems, but we must be careful. For $(-7)^2$ we must begin by evaluating what is in parentheses, but there is nothing to actually do. We then handle the exponent and compute -7 × -7 for an answer of 49.

For -7^2, explain to the student that to make a number negative is, in essence, to multiply it by -1. Since exponents are handled before multiplication, we must first square 7 to get 49. Then, we take that 49 and make it negative by effectively multiplying it by -1. We get an answer of -49.

NOTE TO INSTRUCTOR: There isn't much to practice for this lesson other than problems which are virtually identical to the ones presented. Use any extra time to review the concepts presented in this lesson, and all of the material presented up to this point. The next chapter beings a unit on fractions. There is truly no point in starting that unit until all of the previous material has been mastered. This is especially true for signed number arithmetic. Most students are very quick to proclaim that "fractions are hard," but this is usually because basic arithmetic was never fully mastered.

INSTRUCTOR'S NOTES:

CHAPTER FIVE

Lesson Plans for "Basic Operations with Fractions (+, −, ×, ÷)"

Lessons 21 to 23

Topics Covered in This Chapter:

Introduction to fractions; Adding and subtracting fractions with like (matching) denominators; Multiplying and dividing fractions

MATH MADE A BIT EASIER LESSON PLANS:
A GUIDE FOR TUTORS, PARENTS, AND HOMESCHOOLERS

LESSON 21

AIM/TOPIC(S): Introduction to fractions

CONNECTIONS TO LATER MATERIAL: Obviously fractions play a huge role in later math, especially algebra. The basic concept must be fully mastered at this time. Do not move past this lesson until it is, since later work depends upon it.

WARM-UP / MOTIVATION: Elicit the student's notion of what a fraction is. Ask the student for examples of when fractions are used in everyday life. If the student has already learned about fractions, try to get a sense of how s/he feels about the topic and why. For most students, the study of fractions is the point where they first describe math as "hard."

DEMONSTRATIVE EXAMPLES / POINTS TO ELICIT:
A fraction represents part of a whole. A basic fraction implies a value between 0 (nothing at all), and 1 (the whole thing we are dealing with). In an age-appropriate way, convey the concept of a generic whole. For basic problems it is never a matter of, "What if my pizza pie was larger than yours?" Convey that our generic whole will always be divided into equally-sized parts. It is never a matter of, "What if my slice was thinner than yours because the pizza wasn't cut evenly?"

Draw a shaded pie diagram to represent 3/8. Elicit that the top number (i.e., numerator) represents the number of parts of concern, and the bottom number (i.e., denominator) represents the total number of parts. Then draw a rectangular-based representation the same fraction to stress that we're still dealing with the same fraction, and the same part of a whole.

LESSON PLANS FOR "BASIC OPERATIONS
WITH FRACTIONS (+, −, ×, ÷)"

Elicit that as the denominator of a fraction increases, the value of the fraction decreases since it's representing smaller and smaller portions of a whole. Most students can visualize that 3 slices from a 16-slice pizza pie is less food than 3 slices from a standard 8-slice pie.

Elicit that as the denominator of a fraction decreases, the value of the fraction increases since it's representing larger and larger portions of a whole. Most students can visualize that 3 slices from a pizza which has been cut into 4 slices is more food than 3 slices from a standard 8-slice pie.

Elicit that as the numerator of a fraction increases, the value of the fraction increases since it's representing more parts of a whole. Most students understand that more parts represents a larger portion.

Elicit that as the numerator of a fraction decreases, the value of the fraction decreases since it's representing fewer parts of a whole. Most students understand that fewer parts represent a smaller portion.

PRACTICE EXERCISES: Depending on the age of the student, work with drawings and/or manipulatives (e.g., fraction circles) to drive home the concepts presented. Convey that for our purposes, ½ a bag of candy represents the same numerical value as ½ a pizza pie. It's still the same portion of a whole. Try to explain that in our problems we will be dealing with fractions which are "out of context," and as such, a fraction like ¼ represents a smaller value than ⅞. Explain that a/b is an alternative notation for $\frac{a}{b}$.

INSTRUCTOR'S NOTES:

LESSON 22

AIM/TOPIC(S): Adding/subtracting fractions w/ like denominators

WARM-UP / MOTIVATION: Review the previous lesson. Ask why 1/8 + 2/8 equals 3/8 and not 3/16, and why adding 1/8 + 2/7 is "not that simple." Use this to lead into the lesson.

DEMONSTRATIVE EXAMPLES / POINTS TO ELICIT:
If the student fully internalized the previous lesson, this lesson will be very easy. Stress that when we have like (i.e., matching) denominators, we are dealing with "apples and apples." It's the scenario of, "I ate one slice, and then I ate two more, so I've eaten 3/8 of the pie." Elicit the concept that we must leave the denominator alone, and only add the numerators (i.e., the parts), since it is not the case that our whole suddenly got divided into twice as many parts.

For now, elicit that adding unlike denominators is just "not that simple." Elicit that slices from a pie of 8 slices are smaller than slices from a pie of 7 slices, and as such cannot be simply combined. This topic is addressed in a later lesson.

Elicit that subtraction of fractions with like denominators works as expected. We subtract the numerators and leave the denominator alone. The student will likely be familiar with the idea of reducing his/her answer if possible, but defer this to a later lesson when it is covered.

PRACTICE EXERCISES: The companion workbook offers practice exercises. Optionally supplement with your own.

INSTRUCTOR'S NOTES:

LESSON PLANS FOR "BASIC OPERATIONS
WITH FRACTIONS (+, −, ×, ÷)"

LESSON 23

AIM/TOPIC(S): Multiplying and dividing fractions

WARM-UP / MOTIVATION: Elicit the student's idea of what ½ of ½ equals (use a pizza pie as an analogy). Use this to lead into the lesson. Unfortunately, at this early stage, practical, real-world applications for these operations are hard to find.

DEMONSTRATIVE EXAMPLES / POINTS TO ELICIT:
Revisiting the warm-up, ideally the student will get ¼ after visualizing ½ of ½ of a pie. Explain that in math, "of" means multiplication when appearing between two values.

Show how we can solve the warm-up problem by multiplying $\frac{1}{2} \times \frac{1}{2}$. Explain that the procedure is to multiply straight across the numerators to get the numerator of the answer, and to do the same for the denominators. Explain that this procedure must be memorized.

The first book presented a four-step procedure for dividing a fraction by a fraction. Ex. $\frac{2}{5} \div \frac{3}{7} = \frac{2}{5} \times \frac{7}{3} = \frac{14}{15}$

1) Leave the first fraction alone.
2) Change the division operation to multiplication.
3) Flip the 2nd fraction upside-down to its reciprocal (define).
4) Multiply the two fractions as described above.

PRACTICE EXERCISES: The companion workbook offers practice exercises. Optionally supplement with your own. Reducing/simplifying and "cross-canceling" are covered later.

INSTRUCTOR'S NOTES:

CHAPTER SIX

Lesson Plans for "More About Fractions"

Lessons 24 to 28

Topics Covered in This Chapter:

Four representations of fractions; A fraction is itself a division problem; Multiplying an integer times a fraction; Simplifying a fraction within a fraction; Greatest common factor (GCF); Reducing fractions; Least common multiple (LCM); Adding/subtracting fractions with unlike denominators (LCD)

MATH MADE A BIT EASIER LESSON PLANS:
A GUIDE FOR TUTORS, PARENTS, AND HOMESCHOOLERS

LESSON 24

AIM/TOPIC(S): Four representations of fractions; A fraction is actually a division problem in and of itself

WARM-UP / MOTIVATION: Review and quiz on the previous lessons on fractions

DEMONSTRATIVE EXAMPLES / POINTS TO ELICIT:
Explain that a fraction is actually a division problem in and of itself, namely top divided by bottom. The bar is a dividing line. This is confusing for many students after recently learning how to compute a fraction divided by a fraction. Present the following four equivalent representations of the same fraction:

$$a/b = a \div b = \frac{a}{b} = b\overline{)a}$$

Knowing this, we can deduce that the fraction 7/7 is equal to 1 since $7 \div 7 = 1$. Elicit that any fraction of the form n/n equals 1.

There are times when we need to convert an integer to an equivalent fraction. Elicit that we can do this by putting the integer over a denominator of 1 since any value divided by 1 equals itself. Elicit that any fraction of the form $n/1$ equals n. As an example, we can convert 3 into a fraction as 3/1.

PRACTICE EXERCISES: There is little to practice for this lesson other than to reinforce the above concepts. Use any extra time to review this and all of the previous material on fractions, and to ensure that the student is not harboring any misconceptions or concerns about the topic.

INSTRUCTOR'S NOTES:

LESSON PLANS FOR "MORE ABOUT WITH FRACTIONS"

LESSON 25

AIM/TOPIC(S): Multiplying an integer times a fraction

WARM-UP / MOTIVATION: Ask the student for his/her thoughts on multiplying $3 \times \frac{2}{7}$. Most students will either be uncertain, or will insist that the answer is $\frac{6}{21}$. Ask the student to imagine having eaten 2 slices from a pizza pie which was cut into 7 slices (i.e. 2/7). Now ask the student what fraction would represent three times the portion that s/he ate. Use this to lead into the lesson.

DEMONSTRATIVE EXAMPLES / POINTS TO ELICIT:
Multiplying an integer times a fraction is a very common and easy task, yet one which students get very confused about. The simple rule to remember is that the integer only multiplies the numerator, but let's explore why. Elicit that since we're involved with fractions, we should convert the 3 into a fraction (i.e., 3/1) like we learned in the last lesson. Once that is done we just multiply $\frac{3}{1} \times \frac{2}{7}$ straight across to get 6/7. The integer was always over an "imaginary 1" even if we didn't write it.

Remind the student that we needed to end up with a fraction that is 3 times the value of 2/7. Confirm that 6/7 represents 3 times the value (i.e., the portion) of 2/7.

PRACTICE EXERCISES: There is little to practice for this lesson besides drilling the above concept that the integer only multiplies the numerator of the fraction. Use any extra time to reinforce this lesson, and review/quiz previous material.

INSTRUCTOR'S NOTES:

MATH MADE A BIT EASIER LESSON PLANS:
A GUIDE FOR TUTORS, PARENTS, AND HOMESCHOOLERS

LESSON 26

AIM/TOPIC(S): Simplifying a fraction within a fraction

WARM-UP / MOTIVATION: Ask the student for his/her thoughts on simplifying the fraction shown at right. S/he will likely get an indescribable look on his/her face. Use this to lead into the lesson.
$$\frac{\frac{2}{5}}{\frac{3}{7}}$$

DEMONSTRATIVE EXAMPLES / POINTS TO ELICIT:
Review the lesson in which we learned that a fraction is itself a division problem—top divided by bottom. To simplify the above fraction, we just need to rewrite it as such. We get $\frac{2}{5} \div \frac{3}{7}$ which we solved in Lesson 23. Our fraction happened to have a fraction in the numerator and a fraction in the denominator. Explain that the main fraction bar will always be a bit longer and/or bolder than any other fraction bars in the problem.

Ask the student to simplify the fraction at left. Elicit that we have a fraction in the numerator, and an integer in the denominator. Since we're involved with fractions, we'll convert the integer to a fraction as 3/1. We can rewrite the problem as $\frac{2}{9} \div \frac{3}{1}$ which evaluates to $\frac{2}{27}$ if we follow our procedure for computing a fraction divided by a fraction. Note that our procedure is the same in the case of fractions with an integer in the numerator and a fraction in the denominator.
$$\frac{\frac{2}{9}}{3}$$

PRACTICE EXERCISES: The companion book offers practice problems on this topic. Be certain that the student grasps all of the underlying concepts involved.

INSTRUCTOR'S NOTES:

LESSON PLANS FOR "MORE ABOUT WITH FRACTIONS"

LESSON 27

AIM/TOPIC(S): Greatest common factor (GCF); Reducing/simplifying fractions to lowest terms

CONNECTIONS TO EARLIER / LATER MATERIAL:
This lesson incorporates many earlier concepts including factors and fractions. If those lessons weren't fully mastered, the student will find this lesson very confusing and frustrating. This is one of the most important lessons in the book. If the student doesn't become comfortable using a common factor (ideally the GCF) to reduce a numeric fraction, s/he cannot be expected to later perform the same task with algebraic fractions. Do not move past this lesson until it is fully mastered.

WARM-UP / MOTIVATION:
Review Lesson 10 in which we defined factor by example. Write the numbers 3, 4, 8, 17, 24 and 36, and ask the student to list the factors of each. If s/he is having difficulty, remind him/her of the need to memorize the multiplication table.

Ask the student to examine the factor lists of 24 and 36, and circle the factors which appear on both lists. Then ask him/her to note which of those common factors is the greatest, and define this as the greatest common factor (GCF) of 24 and 36.

Elicit the student's ideas on what we usually do with the fraction 4/8, and how/why. S/he will likely say "½" with uncertainty. If s/he uses the word "reduce" or "simplify," elicit his/her ideas on why 1/2 is a reduction or simplification of 4/8.

Lesson continued on next page →

DEMONSTRATIVE EXAMPLES / POINTS TO ELICIT:

Revisit the warm-up question about 4/8, and the fact that the student had an idea that it was related to 1/2. Elicit that both fractions represent the same value—the same portion of a whole. Use a pizza pie drawing or fraction circles as needed. Elicit the concept that we divided top and bottom of 4/8 by 4 to get 1/2. Many students think in terms of subtraction, and will observe that 3 was subtracted from the numerator, and will ask why we didn't just subtract 3 from the denominator.

This is confusing for many students, but elicit that if we add or subtract the same number from top and bottom, we will end up with a non-equivalent fraction which we never want. We can never change a fraction's actual value.

Elicit that we get from 4/8 to 1/2 by dividing top and bottom by the same value, in this case 4. There is no harm in doing this since we effectively divided the fraction by 4/4 which is 1, and dividing by 1 doesn't change anything.

The student may ask how they would know to divide by 4. Revisit the warm-up task of listing the factors of various numbers. Elicit that any number that we choose to divide by must be a factor of both 4 and 8, otherwise it won't divide into those numbers evenly.

The student will likely ask why we can't use another common factor such as 2. Use this opportunity to elicit that by using the GCF, we can finish our task in just one step, but demonstrate using a factor of 2 twice. Stress that there is no harm in the process taking more than one step, and that this is preferable to wasting valuable time on determining the GCF, especially in the case of fractions involving larger numbers.

LESSON PLANS FOR "MORE ABOUT WITH FRACTIONS"

The student may ask what the point of all this is, and how 1/2 is a "reduced" or "simplified" version of 4/8. Explain that these terms are misnomers to some extent. In math, it is customary to follow the procedure described so that a fraction's numerator and denominator do not have any factors in common other than 1. Using the GCF will result in the process taking only one step, but there is no harm in repeatedly using any common factor until the process is complete.

The student may ask how we know when a fraction is fully reduced. If the GCF of the numerator and denominator is 1, we're done. Examples of reduced fractions are 3/20 and 2/17. Stress that many fractions start out already fully reduced.

Note that some textbooks make use of elaborate and exotic methods for computing the GCF. Most of these are highly impractical in the context of a timed exam, and are very unnecessary. It is rare for a student to encounter a fraction of highly cumbersome numbers, and at worst the student should be able to complete the process in a few steps using a common factor of 2 or 3, even if it isn't the GCF.

PRACTICE EXERCISES: The companion workbook offers practice exercises on this topic, but the student should also practice reducing any fractions that you assign. This is a lesson for which the more practice the better. Before moving ahead, ensure that the student does not struggle with this task at all.

IDEAS FOR EMBELLISHMENT: Revisit the lesson on primes and elicit that if a fraction involves a prime number, there are only two possible values for the GCF: 1 and that number.

INSTRUCTOR'S NOTES:

LESSON 28

AIM/TOPIC(S): Least common multiple (LCM/LCD); Adding and subtracting fractions with unlike denominators

CONNECTIONS TO EARLIER / LATER MATERIAL:
This lesson incorporates many earlier concepts including factors, multiples, and fractions. If those lessons weren't fully mastered, the student will find this lesson very confusing and frustrating. This is among the most important lessons in the book. If the student doesn't become comfortable adding and subtracting fractions with unlike denominators, s/he cannot be expected to later perform the same task with algebraic fractions. Do not move past this lesson until it is fully mastered.

WARM-UP / MOTIVATION:
Ask the student to recite the first 10 multiples of each number between 1 and 12. Elicit that s/he is really just reading across each row of the multiplication table. The student should be able to do this with confidence, in at most two minutes.

Ask the student to add 3/4 + 1/6. Many students will get 4/10 by adding straight across top and bottom. Use pie drawings, fraction circles, or estimation to disprove that answer. Review the lesson on adding/subtracting fractions with like denominators, and elicit that this problem is not as simple as adding across because we don't have common denominators—we have parts of unequal size. Use this idea to lead into the lesson.

DEMONSTRATIVE EXAMPLES / POINTS TO ELICIT:
Revisiting the warm-up, elicit that before we can add the two fractions, we must somehow "convert" each one so that they have like (matching) denominators. Review how we "re-

LESSON PLANS FOR "MORE ABOUT WITH FRACTIONS"

duced" a fraction by dividing top and bottom by the same number, and elicit that in this case we'll do the opposite by multiplying top and bottom by a given number.

Elicit that since we're going to be using multiplication, whatever value we choose as our target common denominator must be a multiple of both denominators.

Have the student extend the lists of multiples of 4 and 6 until there are about 20 values in each list. Elicit that these numbers have many multiples in common (indeed infinite), but that the lowest of those common multiples is 12. We will use 12 as our target common denominator (the LCM), and try to "convert" each fraction so it has 12 as its denominator. Then we will be able to add the two fractions like we've learned.

Let's "convert" the first fraction of 3/4 so that it has a denominator of 12. Stress that at no time will we actually change the value of the fraction. Remind the student that we need to multiply top and bottom by the same number so that the denominator will be 12. That number is 3. We're really multiplying by 3/3 which is equal to 1, which is why this procedure is permitted. 3/4 has been "converted" to 9/12.

Let's "convert" the second fraction of 3/4 so it has a denominator of 12. We need to multiply top and bottom by the same number so that the denominator will be 12. That number is 2. It's OK that we used a different value for this fraction than we did for the first fraction. 1/6 has been "converted" to 2/12. Our modified problem is 9/12 + 2/12 which equals 11/12.

Lesson continued on next page →

Many students will ask what happens if we don't use the LCM as our target common denominator, but instead choose a larger common factor. Demonstrate that this is fine, but after we're done adding or subtracting, we will have to reduce the resulting fraction. In the above example, if we used a common denominator of 24, we would get a sum of 22/24 which would reduce to 11/12 just the same.

See if you can elicit that we can always get a common denominator (but not necessarily the LCD) by multiplying the two involved denominators. In this case, since 4 × 6 is 24, 24 was guaranteed to be a multiple of both 4 and 6. Encourage the student to use this technique if s/he is having difficulty computing the LCD. It is not worth wasting much exam time when it isn't hard to reduce the resulting fraction. Note that many textbooks utilize exotic and unwieldy methods for computing the LCD, all of which are unnecessary in the context of standard exam questions.

At this point, many students start to confuse GCF with LCM. These topics can be tricky. Take whatever time is necessary to review this and the previous lessons until the matter is fully clear. Elicit that for any number, the number itself is both a factor and a multiple of that number.

PRACTICE EXERCISES: The companion workbook offers practice exercises on this topic, but the student should also practice adding/subtracting any fractions with unlike denominators which you assign. This is a lesson for which the more practice the better. Before moving ahead, ensure that the student does not struggle with this task at all.

INSTRUCTOR'S NOTES:

CHAPTER SEVEN

Lesson Plans for "Other Topics in Fractions"

Lessons 29 to 31

Topics Covered in This Chapter:

Mixed numbers; Improper fractions; Converting division problem remainders to fractions; Comparing "cross products" and reducing to lowest terms to determine if two fractions are equivalent; Negative fractions; $0/n$ and $n/0$; "Cross canceling"

MATH MADE A BIT EASIER LESSON PLANS:
A GUIDE FOR TUTORS, PARENTS, AND HOMESCHOOLERS

LESSON 29

AIM/TOPIC(S): Mixed numbers; Improper fractions; Converting division problem remainders to fractions

WARM-UP / MOTIVATION: Elicit the student's ideas about 2¼ and 9/4. Many students have misconceptions about mixed numbers and improper fractions.

DEMONSTRATIVE EXAMPLES / POINTS TO ELICIT:
Elicit that a mixed number is the sum of an integer and a fraction. Addition is implied even though there is no plus sign. Use pie drawings or fraction circles to elicit that the mixed number in the warm-up represents two wholes, and ¼ of a third whole. Reinforce the concept that we are dealing with generic wholes, and that one whole is not larger than the other.

Many students have a hard time grasping improper fractions because they imply a quantity greater than one whole (except those of the form n/n). A student is justified in asking, "How can I possibly order a pizza pie and eat 9/4 of it?"

Review the significance of the numerator and denominator. Nothing has changed. The denominator still tells us how many equal parts our generic whole has been cut into. In the case of 9/4, any fraction circles or pie drawings we work with must be based on fourths (quarters). The numerator tells us how many parts we are dealing with. In this case, we must represent 9 parts. We can start assembling the parts like a jigsaw puzzle.

Demonstrate how we can assemble 4 parts together to make one whole. That took care of 4 of the 9 parts. We still must assemble 5. Elicit that we can't squeeze the 5 parts together to make another whole. The best we can do is use 4 of them to make another whole, and have one part left over. Elicit the equality of 2¼ and the improper fraction 9/4.

Explain that in an improper fraction, the numerator is larger than or equal to the denominator. If they are equal, the fraction is equivalent to 1 which we previously learned. If the numerator is larger, the fraction will represent a value greater than 1 (one whole), which is nothing to be alarmed about.

Usually it is acceptable to leave an answer as an improper fraction unless instructed to convert it to a mixed number. Mathematically, what we did above was divide 9 by 4 to get 2 (two wholes) with 1 left over. Earlier we represented this as 2R1, but we just saw that we can now put the remainder over the original denominator, which in this case is ¼.

To convert a mixed number to an improper fraction, we essentially work in reverse. To convert 2¼ to an improper fraction, we just literally compute 2 + ¼. Convert the 2 to a fraction as 2/1, and then add 2/1 + 1/4 as we learned. We get 9/4. Elicit the "shortcut" in that we really just multiplied the integer times the denominator of the fraction, and then added the numerator of 1 to get the new numerator of 9. The denominator remained 4 since we're still dealing with fourths. In problems involving operations with mixed numbers, it is best to convert each mixed number to an improper fraction.

PRACTICE EXERCISES: The companion workbook offers practice exercises. The student should feel comfortable converting to/from any mixed number or improper fraction.

INSTRUCTOR'S NOTES:

MATH MADE A BIT EASIER LESSON PLANS:
A GUIDE FOR TUTORS, PARENTS, AND HOMESCHOOLERS

LESSON 30

AIM/TOPIC(S): Comparing "cross products" and reducing to lowest terms to determine if two fractions are equivalent

WARM-UP / MOTIVATION: Elicit the student's thoughts on the difference between 4/6 and 14/21. Then ask the student to reduce each fraction to lowest terms.

DEMONSTRATIVE EXAMPLES / POINTS TO ELICIT:
Revisit the warm-up, and note that since both fractions reduce to 2/3, it means that the two are equivalent. This is one of two methods we can use to determine if two fractions are equivalent. If they reduce to the same value then they are.

The other method involves comparing the "cross products" to see if they are equal. Let's compare $\frac{2}{5}$ to $\frac{4}{11}$. Informally, the cross products are formed by multiplying across the diagonals. This is sometimes referred to as "cross multiplying." In this example, one cross product is 2×11 and the other is 5×4. The cross products are not equal. This proves that the two fractions are not equivalent. For practice, compare the cross products of the warm-up problem fractions to see that they are equal.

Stress that "cross multiplying" is not the same as the "multiplying across" that we do when multiplying fractions. "Cross multiplying" is what we do to compare the cross products of two fractions to see if they are equivalent.

PRACTICE EXERCISES: The companion workbook offers exercises for extra practice with these concepts.

INSTRUCTOR'S NOTES:

LESSON PLANS FOR "OTHER TOPICS IN FRACTIONS"

LESSON 31

AIM/TOPIC(S): Negative fractions; $0/n$; $n/0$; "Cross canceling"

WARM-UP / MOTIVATION: This lesson reviews many previous concepts. Use the warm-up segment to review any concepts which the student seems insecure about.

DEMONSTRATIVE EXAMPLES / POINTS TO ELICIT:
Ask the student to compare -2/3, 2/-3, and -2/-3. We must remember the rules for signed number division, and that a fraction is actually a division problem. In the first two fractions we are dividing values of opposite signs which results in a negative. It doesn't matter that one fraction had the negative on top and one had it on the bottom. Both evaluate to $-\frac{2}{3}$. Call attention to this alternate way of representing negative 2/3 by writing the negative sign beside the fraction bar.

In the third fraction we have a negative divided by a negative which is positive. Informally, the negatives "cancel out," and we are left with positive ⅔.

Elicit the student's ideas on the fraction 0/8. Review the significance of the numerator and denominator. We are dealing with 0 parts. It doesn't matter how many parts our whole was cut into. The fraction is equal to 0. $0/n = 0$

Elicit the student's ideas on the fraction 7/0. This is a bit harder for students to grasp, and there are many different ways of explaining it. One effective way is to elicit the student's concept of 7/10000, followed by 7/1000, 7/100, 7/10, 7/1, etc. If the student can handle it, continue the pattern with fractional

79

denominators which get smaller and smaller. Elicit that as the denominator gets smaller, the size of our portion increases. As the denominator approaches 0, the value of the fraction approaches infinity. We define this scenario as "Undefined." We are "not allowed" to divide by 0. Demonstrate that a calculator will generate an error message if you try.

Elicit the student's thoughts on multiplying $\frac{5}{12} \times \frac{8}{9}$. Most students are familiar with and have no difficulty using the "cross canceling" technique of dividing the 8 and 12 by a common factor of 4, resulting in 2 and 3, respectively. Elicit that this can only be done when multiplying two fractions, and not with any other operation. Also elicit that we are allowed to do this because the numerators will get multiplied together anyway, as will the denominators. If we don't first "pull out" common factors from the tops and bottom, we'll just have the extra step of reducing the answer later.

PRACTICE EXERCISES / POINTS TO REEMPHASIZE:

The companion workbook offers exercises for extra practice with these concepts. Reemphasize that we are never allowed to divide by 0. Any answer other than the word "undefined" is wrong. Take extra time to review what was taught about negative fractions. Explain that much of what is being learned now will be revisited in the study of algebra. Some of it will also make more sense and seem more practical at that time.

Since this lesson is short and rather simple, use any extra time to thoroughly review and quiz on all the material to this point. After doing a bit more with fractions in the next chapter, we will be moving on to decimals and percents.

INSTRUCTOR'S NOTES:

CHAPTER EIGHT

Lesson Plans for "The Metric System, Unit Conversion, Proportions, Rates, Ratios, and Scale"

Lessons 32 to 36

Topics Covered in This Chapter:

Non-metric units; Time spans; Temperature; Metric units; Metric prefixes; (Unit) Ratios and Rates; Cost per unit; Converting measurements using unit ratios; Solving basic proportions; Scale drawings

MATH MADE A BIT EASIER LESSON PLANS:
A GUIDE FOR TUTORS, PARENTS, AND HOMESCHOOLERS

LESSON 32

AIM/TOPIC(S): Non-metric units; Time spans; Temperature

WARM-UP / MOTIVATION: Survey the student to see how much lesson content s/he already knows. This lesson may be skipped if coursework or upcoming exams do not require it.

POINTS TO MEMORIZE: There are 12 inches in a foot, 3 feet in a yard, and 5280 feet in a mile. There are 16 ounces in a pound, and 2000 pounds in a (US) ton. A cup is a unit of volume (capacity), not weight (mass). A cup is defined as 8 fluid ounces which can be confusing since the non-fluid ounce is a unit of weight (mass). There are 2 cups in a pint, and 2 pints in a quart. There are 4 quarts in a gallon.

There are 60 seconds in a minute, and 60 minutes in an hour. Fractions involving time have a denominator of 60, and are then reduced. There are 365/366 days in a non-leap/leap year. A week has 7 days, and a year has 52 weeks or 12 months.

Converting between °C and °F is usually saved for algebra. Ensure that the student won't panic if a problem involves temperatures in °C. Note that water freezes/boils at 0/100 °C.

IDEAS FOR EMBELLISHMENT: There are many ways that this lesson can be brought to life, but unless the student is quite young, lesson time is much better spent on other topics.

PRACTICE EXERCISES: The companion workbook offers exercises for extra practice with these concepts.

INSTRUCTOR'S NOTES:

LESSON PLANS FOR "THE METRIC SYSTEM, UNIT CONVERSION, PROPORTIONS, RATES, RATIOS, SCALE"

LESSON 33

AIM/TOPIC(S): Metric units; Metric prefixes

WARM-UP / MOTIVATION: Survey the student to see how much lesson content s/he already knows. This lesson may be skipped if coursework or upcoming exams do not require it.

POINTS TO MEMORIZE:
The basic unit of volume (capacity) is the liter ($\approx$ one quart). The basic unit of mass (weight) is the gram (30 g $\approx$ 1 oz.). The basic unit of length is the meter ($\approx$ one yard).

The prefix "kilo-" means to multiply the unit times 1000. A kilogram is 1000 grams ($\approx$ 2.2 pounds). The prefix "milli-" means to divide the unit by 1000. A milliliter is 1/1000 of a liter ($\approx$ 20 drops of water). The prefix "centi-" means to divide the unit by 100. A centimeter is 1/100 of a meter ($\approx$ ½ an inch).

These are by far the most common prefixes. It is unlikely that the student will be responsible for knowing any others, but obviously confirm this and instruct accordingly. A later lesson covers how to convert from one unit to another.

IDEAS FOR EMBELLISHMENT: There are many ways that this lesson can be brought to life, but unless the student's coursework or math goals involve this material, lesson time is much better spent on other topics.

PRACTICE EXERCISES: The companion workbook offers exercises for extra practice with these concepts.

INSTRUCTOR'S NOTES:

LESSON 34

AIM/TOPIC(S): (Unit) Ratios and Rates; Cost per unit

NOTE TO INSTRUCTOR: In upcoming lessons we will work with unit conversion and proportions.

WARM-UP / MOTIVATION: Present the following problem: "A school has 4368 students and 127 teachers. What is the student/teacher ratio?" Then present this problem: "A bushel of 42 apples costs $37. At that rate, what is the cost of 1 apple?" Use any responses to lead into the lesson.

DEMONSTRATIVE EXAMPLES / POINTS TO ELICIT:
Revisit the first warm-up question. Define a ratio as a comparison of two quantities which have the same units, in this case, people. Elicit that a ratio is effectively a fraction, which in turn is a division problem. The ratio "a to b" can also be represented as $a:b$, a/b, or $\frac{a}{b}$, all of which mean $a \div b$.

While we can represent this ratio as 4368/127, the question implies that we must compute how many students, on average, are assigned to each teacher. To do this we must get the denominator (which represents teachers) to be 1. Elicit that we can do this by dividing top and bottom by 127, which is effectively to just do the computation $4368 \div 127$. We get a student to teacher ratio of 34.4 (rounded) to 1.

Present this problem: "An urn has 3 red marbles and 5 blue marbles. What is the ratio of blue marbles to total marbles?" Elicit that the answer is 5:8, which must be presented in that order and not reversed. Elicit that 3 is not part of the answer.

LESSON PLANS FOR "THE METRIC SYSTEM, UNIT CONVERSION, PROPORTIONS, RATES, RATIOS, SCALE"

A ratio does not necessarily have to be reduced or simplified to have a denominator of 1. For example, if a party is attended by 90 women and 60 men, the female to male ratio is best expressed as 3:2. That paints a better picture than 1.5 to 1. A problem's context and instructions will make the matter clear.

Revisit the second warm-up problem. Define a rate as being similar to a ratio except it deals with a comparison of quantities which have unrelated units, in this case apples and dollars. Elicit that another common rate is miles per hour which is a comparison of distance and time.

Since the problem is requesting the price of one apple, we will have to convert the rate into what is called a unit rate with a denominator of 1. Elicit that the word unit implies "one." While it might be tempting to write 42/37, that is backwards. The dollar amount must come first since we want to know the cost per unit. The order of the values must match. Demonstrate including the units in the fraction (e.g., $\frac{37\ dollars}{42\ apples}$). Since we want to the denominator to be 1, we can divide top and bottom by 42 which is effectively the same as computing 37 ÷ 42. We get a cost of $0.88 (88 cents) per (one) apple. Elicit that we typically round to the nearest penny.

The most common type of unit rate problem involves computing the cost per unit which is done by taking the total cost and dividing it by the total number of items involved. Care must be taken to not perform that task backwards.

PRACTICE EXERCISES: The companion workbook offers exercises for extra practice with these concepts.

INSTRUCTOR'S NOTES:

MATH MADE A BIT EASIER LESSON PLANS:
A GUIDE FOR TUTORS, PARENTS, AND HOMESCHOOLERS

LESSON 35

AIM/TOPIC(S): Converting measurements using unit ratios

WARM-UP / MOTIVATION: Review the previous lesson on ratios and rates. Ask the student to convert 60 feet to inches. Then ask the student to convert 1500 milliliters to liters. Use any responses to lead into the lesson.

DEMONSTRATIVE EXAMPLES / POINTS TO ELICIT:
Revisit the first warm-up question. Most students instinctively know that they must do something involving the number 12, but are confused about precisely what to do. Many students assume that they must divide 60 by 12 to get answer of 5 which is wrong. Explain that while it is easy to get confused with problems involving unit conversion, you about to present a method which will make these problems simple.

For the first problem, elicit that we want the unit of feet to "go away," and we want to replace it with a unit of inches. Elicit that there are 12 inches in 1 foot, and that the ratio of $\frac{1\,foot}{12\,inches}$ is equal to 1 since the top and bottom values are equal. It is no different than any fraction of the form n/n. Remind the student that there is never any harm in multiplying a value times 1 (in any form), since doing so doesn't change the value.

Our plan is to take 60 feet (over an imaginary 1) and multiply it by some version of the above fraction such that the unit of feet "cancels out," and the unit of inches remain. Let's try setting up the problem as $\frac{60\,foot}{1} \times \frac{1\,foot}{12\,inches}$. Elicit that by setting up our problem like this, the unit of feet will not "cancel" since it appears on the top in both fractions. Most students feel OK with this idea since they're used to "cancelling" common factors when multiplying numeric fractions. Elicit that names of units can be "cancelled" in the same way.

Now let's use the reciprocal of our unit ratio. Elicit that the reciprocal is still equal to 1. Let's compute $\frac{60 \; \cancel{feet}}{1} \times \frac{12 \; inches}{1 \; \cancel{foot}}$. The units of feet "cancel" out, and we are only left with inches which is what we want. Multiply straight across to get 720 inches (omitting the denominator of 1 which doesn't matter).

Revisit the second warm-up problem. Elicit that it can be solved by remembering that there are 1000 mL in 1 L, and by setting up our expression as $\frac{1500 \; \cancel{mL}}{1} \times \frac{1 \; L}{1000 \; \cancel{mL}}$. The units of milliliters "cancel," leaving us with only liters. Multiply straight across to get 1500/1000 L which is 1.5 L after dividing.

PRACTICE EXERCISES: The companion workbook offers exercises for extra practice with these concepts. This topic is very popular on standardized exams, and is a topic that has many everyday practical applications. It is also an easy topic as long as the procedures are followed, and basic conversion factors are memorized. Furthermore, this topic serves to review many important concepts which have been covered. For all of these reasons, spend a bit of extra time on this topic.

POINTS TO REEMPHASIZE: Despite how basic this topic is, many students insist on guessing answers, falling victim to decoy answers such as 5 in the first warm-up problem. Spend extra time drilling, and insisting that the student take care in choosing how to set up the unit ratio in each problem.

INSTRUCTOR'S NOTES:

MATH MADE A BIT EASIER LESSON PLANS:
A GUIDE FOR TUTORS, PARENTS, AND HOMESCHOOLERS

LESSON 36

AIM/TOPIC(S): Solving basic proportions; Scale drawings

WARM-UP / MOTIVATION: Elicit the student's ideas on the problem $\frac{1}{5} = \frac{4}{?}$. If s/he gets the correct answer, ask if there was another way the problem could have been solved. Assess the student's ability to interpret scale drawings and map legends. Use any responses to lead into the lesson.

DEMONSTRATIVE EXAMPLES / POINTS TO ELICIT:
Explain that a proportion is a way of showing that two ratios (effectively fractions) are equivalent. Most students intuitively feel comfortable with this concept. Revisit the warm-up question. Elicit that the numerator of the second fraction is 4 times the numerator of the first fraction. That means that the unknown denominator will have to be 4 times the denominator of the first fraction if we want the second fraction to be equivalent. The missing value is 20. Elicit that we can also apply the logic that since the denominator of the first fraction is 5 times its numerator, the second fraction will have to parallel that relationship. Using this method we also get 20.

Until the study of algebra, problems involving proportions will be no more complicated than this one. Elicit that we can check our answer by reducing both fractions to lowest terms to confirm that both fractions are equal. We can also use the "cross-multiplying" technique from Lesson 30 to confirm that the cross products of the two fractions are equal.

Present some maps and/or scale drawings with legends like the one shown. It is not worth spending much time on this topic for two reasons. The first is that most students don't have any difficulty with it as long as they feel comfortable with

proportions in general. The second reason is because no matter how many problems you practice, a problem encountered on an exam will likely look different. A student who is uncomfortable with the underlying concept of proportions will just blame that "the picture was different."

A scale drawing or map problem is no different than any basic proportion problem. We are given three out of four pieces of information, and we need to determine the fourth. The only tricky part is correctly translating the picture into a proportion. Elicit that there are two ways we can set up the above problem to determine the actual length of the bicycle. One way is $\frac{1\,in.}{3\,in.} = \frac{2\,ft.}{?\,ft}$. Another way is $\frac{1\,in.}{2\,ft.} = \frac{3\,in.}{?\,ft}$. Elicit that both methods will give us the correct answer of 6 ft, and that in each case we are comparing "apples to apples" but in different ways.

PRACTICE EXERCISES: The companion workbook offers exercises for extra practice with these concepts.

POINTS TO REEMPHASIZE: Elicit that in a proportion, we do our comparisons by way of multiplication or division. For example, we could say that one denominator is ½ or 3 times the size of the other. Proportions are never solved in terms of addition or subtraction. Elicit that 5/6 does not equal 7/8. Many students will erroneously apply the logic of "+1" when comparing numerators to denominators, or "+2" when comparing across.

INSTRUCTOR'S NOTES:

CHAPTER NINE

Lesson Plans for "Working with Decimals"

Lessons 37 to 42

Topics Covered in This Chapter:

Introduction to decimals; Adding and subtracting decimals; Comparing decimals including those commonly confused; Writing and saying decimal numbers; Converting decimals to fractions

LESSON 37

AIM/TOPIC(S): Introduction to decimals (Note: Later lessons cover operations and other topics involving decimals)

WARM-UP / MOTIVATION: Survey the student to assess his/her comfort level with decimal numbers and place value. Elicit his/her ideas on the value 0.5. If s/he says "½," ask what 2 has to do with 5. Use any responses to lead into the lesson.

DEMONSTRATIVE EXAMPLES / POINTS TO ELICIT:
Most students have a general sense of what a decimal number is just from their experience with money (dimes and pennies). Elicit that the decimal place values represent part of a whole. Review the place value chart for whole numbers, and elicit that each place is 1/10 the value of the place on its left. This pattern continues to the right of the decimal point which is used to separate the whole and fractional number place values.

Elicit the decimal place values through at least ten thousandths. Elicit that each ends with "-ths," and that care must be taken to not confuse decimal places values with the similar-sounding whole number counterparts. Elicit why there is no "oneths" place. Reinforce the concept of our base-10 system in which every place is 10 times as big as the place on its right. Note that we don't use commas in the decimal place values.

Revisit the warm-up question about 0.5. Elicit that since the 5 is in the tenths place it represents 5/10 which reduces to ½. Try to insist that such a value be read as "five tenths" as opposed to the more common "point five."

INSTRUCTOR'S NOTES:

LESSON PLANS FOR "WORKING WITH DECIMALS"

LESSON 38

AIM/TOPIC(S): Adding and subtracting decimals

WARM-UP / MOTIVATION: Review adding and subtracting two-digit whole numbers. Ask the student to try adding $27.89 + $6.43. Then try $81.93 − $25.14. Then try 23.4 + 56.78. If the student has trouble, just elicit his/her ideas. Explain the importance of being able to do basic calculations like this by hand even if a calculator is permitted on exams and in class.

DEMONSTRATIVE EXAMPLES / POINTS TO ELICIT:
If the student is confident with adding and subtracting whole numbers, s/he should not have any difficult adding or subtracting dollar amounts, and will likely intuitively know to line up the decimal points and place values. Of course explain the procedure if needed, eliciting the logic behind lining up the decimal place values as we lined up whole number places.

Sometimes confusion can set in when adding or subtracting decimal numbers which do not contain the same number of decimal digits. Emphasize that we always line up the decimal points, as well as all place values (tenths over tenths, hundredths over hundredths, etc.) Zeroes may be added to fill empty places as needed. For example, in the third warm-up problem, the first addend may be written as 23.40.

PRACTICE EXERCISES: The companion workbook offers exercises for extra practice with these concepts. In the unlikely event that the student is responsible for multiplying and dividing decimals by hand, supplement accordingly.

INSTRUCTOR'S NOTES:

LESSON 39

AIM/TOPIC(S): Comparing decimal numbers which are commonly confused or misunderstood

WARM-UP / MOTIVATION: Review all the previous material on decimals. Elicit the student's ideas on these comparisons: 1) 2.3 vs. 2.03; 2) 9.3 vs. 9.30; 3) 0.5 vs. .5; 4) 17 vs. 17.0

DEMONSTRATIVE EXAMPLES / POINTS TO ELICIT:
Revisit the first warm-up problem. Elicit that both numbers have 2 wholes, but the first also has 3 tenths, while the second has 3 hundredths (and 0 tenths) which is worth less. Elicit how the 0 is serving as a placeholder, and that we can't omit it.

Revisit the second warm-up problem. Elicit that both numbers have 9 wholes and 3 tenths. The value of the second number is not changed by tacking on 0 hundredths. Such a non-significant digit is usually dropped unless we are dealing with money, or to specifically show that nothing is in that place.

Revisit the third warm-up problem. Elicit that both numbers have 5 tenths and no wholes. The first version is preferable in print because it makes this fact easy to see.

Revisit the fourth warm-up problem. Elicit that both numbers have 17 wholes and no decimal component. The value of the second number is not changed by tacking on 0 tenths. Such a non-significant digit is usually dropped unless we are dealing with money (e.g., $17.00), or have some other specific reason.

INSTRUCTOR'S NOTES:

LESSON PLANS FOR "WORKING WITH DECIMALS"

LESSON 40

AIM/TOPIC(S): Writing and saying decimal numbers

WARM-UP / MOTIVATION: Elicit the student's ideas on how to say these decimal numbers: 1) 203.1; 2) 5.37 3) 0.902. Use any responses to lead into the lesson.

DEMONSTRATIVE EXAMPLES / POINTS TO ELICIT:
Explain that whole and decimal numbers are sometimes spelled in words, and that if the student is unable to properly interpret them, s/he will get easy exam questions wrong.

The main point to emphasize is that we use the word "and" to represent a decimal point, and we do not use the word "and" for any other purpose. The number 101 is "one hundred one." There is no "and" regardless of any spotted-dog movie.

Revisit the warm-up problems. The first number is read as "two hundred three and one tenth." Try to break the student's habit of saying "point one." The second number is read as "five and thirty-seven hundredths." We look how far to the right the decimal digits reach, and we use that place value's name in what we say or write. The third number is read as "nine hundred two thousandths." We omit "zero and" since there is only a decimal component. We then treat the decimal digits almost as though they are a whole number, and say that number followed by the decimal place to which they extend.

PRACTICE EXERCISES: The companion workbook offers exercises for extra practice with this topic.

INSTRUCTOR'S NOTES:

MATH MADE A BIT EASIER LESSON PLANS:
A GUIDE FOR TUTORS, PARENTS, AND HOMESCHOOLERS

LESSON 41

AIM/TOPIC(S): Comparing decimal numbers

WARM-UP / MOTIVATION: Review all the previous material on decimals. Elicit the student's ideas on these comparisons: 1) 23.4 vs. 8.5679; 2) 0.489 vs. 0.6; 3) 0.5137 vs. 0.512999

DEMONSTRATIVE EXAMPLES / POINTS TO ELICIT:
Revisit the first warm-up problem. Elicit that the whole number portion of a number always takes precedence over the decimal portion. In the first example, 23 is greater than 8. We are done. The decimal portions of the numbers do not matter.

In the second problem, both numbers have no whole number component, so we must move to the right to continue our comparison. The first number has 4 tenths, and the second number has 6 tenths. This means that the second number is bigger. We are done. Ensure that the student does not erroneously think that "489" is bigger than "6." Those digits are serving as decimal place values. It doesn't matter that we tacked on 8 hundredths and 9 thousandths to the first number. It still cannot make it bigger than 6 tenths.

Revisit the third warm-up problem. We start our comparison on the left, and work our way to the right to "break ties." 5 is tied with 5, and 1 is tied with 1. Now we're up to the thousandths place where 3 is bigger than 2, and we are done.

PRACTICE EXERCISES: The companion workbook offers exercises for extra practice with this topic.

INSTRUCTOR'S NOTES:

LESSON PLANS FOR "WORKING WITH DECIMALS"

LESSON 42

AIM/TOPIC(S): Converting decimals to fractions

NOTE TO INSTRUCTOR: This is a short and simple lesson so use the warm-up time to completely review all of the lessons on decimals. Ensure that the student doesn't view decimals as some mystical, isolated topic in math, and that s/he isn't intimidated by them. Take extra time to review the lesson on reading and writing decimal numbers since that ties directly in to the topic of this lesson.

DEMONSTRATIVE EXAMPLES / POINTS TO ELICIT:
Let's convert 0.3 into a fraction. We read that number as "three tenths," and that is how we write it as a fraction: 3/10.

Let's convert 0.47 into a fraction. We read that number as "forty-seven hundredths," and that is how we write it as a fraction: 47/100.

Let's try one more and convert 0.505 into a fraction. We read that number as "five hundred five thousandths," and that is how we write it as a fraction: 505/1000.

That is all there is to converting decimals into fractions, other than reducing the resulting fractions to lowest terms if possible. Elicit that our place value system makes the job easy since we already know a decimal's value when we see it.

PRACTICE EXERCISES: The companion workbook offers exercises for extra practice with this topic.

INSTRUCTOR'S NOTES:

CHAPTER NINE AND FIVE-TENTHS

Lesson Plans for "More Topics in Decimals"

Lessons 43 to 48

Topics Covered in This Chapter:

Rounding decimals; Multiplying/dividing by powers of 10; Converting fractions to decimals (two methods); Repeating and terminating decimals; Decimals within fractions; Comparing fractions without converting to decimals; Scientific notation

LESSON 43

AIM/TOPIC(S): Rounding decimals; Multiplying and dividing by powers of 10

NOTE TO INSTRUCTOR: Use the warm-up time to quiz and review the lessons on decimal place values and rounding whole numbers. This lesson is just a combination of those concepts, so don't begin it until those concepts are mastered.

DEMONSTRATIVE EXAMPLES / POINTS TO ELICIT:
There is very little to demonstrate and elicit for rounding decimals. Practice rounding decimal numbers to various places using the exact same procedure that we used to round integers. Ensure that the student carefully reads all instructions. Even if a number has a decimal component, we may still be instructed to round it to a whole number place.

Demonstrate the "shortcuts" for multiplying and dividing values by powers of 10 by moving the decimal point right or left a given number of "hops" (10=1, 100=2, 1000=3, etc.), and inserting placeholder zeroes as needed. Elicit that whole numbers have an "invisible" decimal point on the right.

Surprisingly, many students are uncomfortable using these shortcuts even though they are quick and easy. Many students aren't convinced that the shortcuts always work, and prefer to compute such problems by hand or with a calculator.

PRACTICE EXERCISES: The companion workbook offers exercises for extra practice with these topics.

INSTRUCTOR'S NOTES:

LESSON PLANS FOR "MORE TOPICS IN DECIMALS"

LESSON 44

AIM/TOPIC(S): Converting fractions to decimals

NOTE TO INSTRUCTOR: This is a short and simple lesson so use the warm-up time to completely review all of the lessons on decimals. Review that a fraction is a division problem in and of itself—top divided by bottom.

DEMONSTRATIVE EXAMPLES / POINTS TO ELICIT:
To convert a fraction to a decimal, all we must do is compute numerator divided by denominator, in that order. It is very common for students to reverse the order because they are not comfortable with a computation such as $1 \div 2$.

This book assumes that the student is permitted to a calculator for such computations which is very likely. Obviously teach the long division if necessary.

POINTS TO REEMPHASIZE: Ensure that the student computes top divided by bottom. Many students have a hard time accepting this. Some think that the order doesn't matter, or they key it in wrong into their calculator. Have the student estimate the decimal value of common fractions to convince him/her that the procedure works correctly.

PRACTICE EXERCISES: Practice converting various fractions to decimals, rounding to the nearest tenth or hundredth. In upcoming lessons we will learn to categorize the resulting decimals as repeating or terminating. The companion workbook offers additional exercises on this topic.

INSTRUCTOR'S NOTES:

LESSON 45

AIM/TOPIC(S): Converting fractions to decimals when the fraction can be altered to have a power-of-10 denominator

WARM-UP / MOTIVATION: Ask the student to convert fraction 7/50 to a decimal using the method from the previous lesson. Then ask if there is any way to get the answer without having to compute top divided by bottom. Offer the hint of altering the fraction so that it has a power-of-10 denominator.

DEMONSTRATIVE EXAMPLES / POINTS TO ELICIT:
If necessary, remind the student of what a power of 10 is (10, 100, 1000, etc.) Revisit the warm-up problem, and elicit that there is no harm in multiplying top and bottom by 2 since doing so is to effectively multiply by 2/2 which is 1. We now have 14/100, which can just be written as 0.14.

Elicit that whenever it is easy to alter a fraction so that it has a denominator which is a power of 10, we can see right away what its decimal equivalent is. Elicit that sometimes we will divide top and bottom by the same number. For example, to convert 46/2000 to a decimal, we could divide top and bottom by 2 to get 23/1000 which is 0.023.

Elicit that this technique should only be used when there is some obvious way to alter the fraction so that it has a power-of-10 denominator. In all other cases, use the previous method.

PRACTICE EXERCISES: The companion workbook offers exercises for extra practice with this topic.

INSTRUCTOR'S NOTES:

LESSON PLANS FOR "MORE TOPICS IN DECIMALS"

LESSON 46

AIM/TOPIC(S): Repeating and terminating decimals; Decimals within fractions

WARM-UP / MOTIVATION: Ask the student to convert these fractions into decimals by using a calculator to compute top divided by bottom: 3/10; 2/3; 5/7. Have the student examine the results carefully.

DEMONSTRATIVE EXAMPLES / POINTS TO ELICIT:
Revisit the first warm-up problem. The calculator returns an answer of 0.3. Elicit that this can be described as a terminating decimal—it stops at the 3. If we computed 3 ÷ 10 using long division, we would get 0.3000..., but all the zeroes to the right of the 3 are insignificant and therefore dropped.

Elicit that any fraction which has a power-of-10 denominator or which can be altered to have one will convert to a terminating decimal. Some examples are 17/25, 9/500, and 11/40

Revisit the second warm-up problem. The calculator returns something like 0.66666667. Many students have a hard time accepting that the sixes actually repeat forever. If necessary, demonstrate with long division to prove the point. Elicit that this can be described as a repeating decimal. We write $0.\bar{6}$ (note the bar) to show that the sixes continue forever. This is not the same as 0.6 which converts to 6/10.

Demonstrate that some fractions convert to decimals which have some but not all of their digits repeat. A simple example is 5/6 which converts to 0.8333... which is written as $0.8\bar{3}$.

Revisit the third warm-up problem. The calculator returns 0.714285... with the pattern repeating. It may be necessary to use a computer's calculator to fully see the pattern. Elicit that this is also a repeating decimal since the pattern repeats infinitely. We write it as $0.\overline{714285}$ to show the repeating digits.

On a different topic, elicit the student's ideas about the fraction 0.7/23. Many students will ask, "Is it a fraction or a decimal?" Elicit that it is a fraction which happens to have a decimal in the numerator. This is not of concern. When working with such fractions, if desired, the fraction can be "simplified" by multiplying top and bottom by a power of 10 so as to "clear" the decimal.

As an example, in the fraction 0.7/23, we can multiply top and bottom by 10 to get the equivalent 7/23. For the fraction 0.87/49, we can multiply top and bottom by 100 to get the equivalent 87/4900.

IDEAS FOR EMBELLISHMENT: Most students are amused by the decimal equivalents of fractions such as 5/9, 34/99, 123/999, etc., so consider demonstrating them.

Optionally, explain that in addition to repeating and terminating decimals, there is a concept of a non-repeating decimal, a common example of which is pi (π). Obviously this will take the lesson off on a tangent which may not be desired. Pi is covered in the algebra/geometry books of this series.

PRACTICE EXERCISES: The companion workbook offers exercises for extra practice with this topic.

INSTRUCTOR'S NOTES:

LESSON PLANS FOR "MORE TOPICS IN DECIMALS"

LESSON 47

AIM/TOPIC(S): Comparing fractions w/o converting to decimals

WARM-UP / MOTIVATION: Ask the student to visually determine which of these fractions is bigger: 137/413 or 529/1590. Elicit why this task is hard. Repeat this task for these fractions: 503/2134 or 16/21. Elicit why this is easy.

DEMONSTRATIVE EXAMPLES / POINTS TO ELICIT:
Remind the student that we can always convert fractions to decimals for easy comparison. Revisit the first warm-up comparison. The fractions convert to 0.3317 and 0.3327 (rounded), respectively, proving that the second one is bigger. We really needed to do the computation in order to tell.

Elicit that in many cases we can solve such problems visually, without the need for computation. Revisit the second warm-up comparison. Elicit that the first fraction visually appears to be about ¼. The second fraction is visibly greater than ⅔, and is certainly bigger than ½, both of which are bigger than ¼. That is all we need to know. The second fraction is definitely bigger.

Whenever we are comparing fractions, we should take a moment to determine if the fractions are comparable to common "reference fractions" such as ½ and ¼. If they are, it is not necessary to do any decimal conversion. It takes practice to gain experience with making this determination.

PRACTICE EXERCISES: The companion workbook offers exercises for extra practice with this topic.

INSTRUCTOR'S NOTES:

LESSON 48

AIM/TOPIC(S): Scientific notation

WARM-UP / MOTIVATION: Elicit the student's ideas about when we might see numbers such as 4,370,000,000 and 0.00000003504, and why they are cumbersome.

DEMONSTRATIVE EXAMPLES / POINTS TO ELICIT:
Elicit that in the above numbers, all we really care about are the significant digits and how many zeroes are involved. Scientific notation is a shortcut for this purpose. It is often used when writing very large numbers like interstellar distances, and very small numbers like microscopic measurements.

Numbers in scientific notation are represented in the form $a \times 10^b$. The a value must be a value that is greater than or equal to 1, and less than 10. It is where we put the significant digits of the value that we are representing. In the first warm-up example, the significant digits are 437, which we must write as 4.37 if we want it to be between 1 and 10 as required.

The b value tells us how many places we must move the decimal point left (negative) or right (positive) in order to get back our original number. In the first example, the b value is 9, making the scientific notation 4.37×10^9. Have the student confirm that the second value is 3.504×10^{-8} in scientific notation. Elicit that the 0 in between the 5 and 4 is significant.

PRACTICE EXERCISES: The companion workbook offers exercises for extra practice with this topic.

INSTRUCTOR'S NOTES:

CHAPTER TEN

Lesson Plans for "Working with Percents"

Lessons 49 to 55

Topics Covered in This Chapter:

Introduction to percents; Converting from decimals and fractions to percents; Percents less than 1% and greater than 100%; Percent of increase/decrease; Percent "of" vs. "off"; Increasing/decreasing a value by a given percent; Sales tax; Two common models of word problems; Common equivalent fractions, decimals, and percents

LESSON 49

AIM/TOPIC(S): Introduction to percents

WARM-UP / MOTIVATION: Survey the student to get his/her ideas about what a percent is, and where they are commonly seen and used. Use any response to lead into the lesson.

DEMONSTRATIVE EXAMPLES / POINTS TO ELICIT:
Most students have an idea that a percent represents part of a whole, but don't realize that we can easily convert such a value to a decimal or fraction. A percent is just a simple and concise way of representing part of a whole.

Elicit that the word "percent" literally means "out of 100." This means that $n\%$ is simply $n/100$. Many students cannot accept that the matter is that simple. If we want to convert a percent to a fraction, just drop the % sign and put the value over 100, then reduce the resulting fraction if possible.

Once we have converted a percent to a fraction, it is easy to convert it to a decimal by computing top divided by bottom. This is nothing more than dividing by 100, the shortcut being to move the decimal point two places to the left like we learned. An example is $37\% = 37/100 = 0.37$. Explain that if we are going to do any type of computation involving a percent, it is usually best to first convert the percent to a decimal.

PRACTICE EXERCISES: The companion workbook offers exercises for extra practice with this topic.

INSTRUCTOR'S NOTES:

LESSON PLANS FOR "WORKING WITH PERCENTS"

LESSON 50

AIM/TOPIC(S): Converting from decimals and fractions to percents

WARM-UP / MOTIVATION: Review the previous lesson, then try to elicit the process for converting decimal to a percent.

DEMONSTRATIVE EXAMPLES / POINTS TO ELICIT:
The process to convert a decimal to a percent is the exact opposite of converting a percent to a decimal. For example, to convert 40% to a decimal, we drop the percent sign and move the decimal point two places to the left, resulting in 0.40 which becomes 0.4. Elicit the implied intermediate step of converting 40% into 40/100, and then computing 40 ÷ 100.

Having just done this, elicit that we can convert 0.4 to a percent by reversing the steps. First we move the decimal two places to the right, and then we add a % sign to get 40%.

One way to convert a fraction to a percent is to alter the fraction to have a denominator of 100 if possible. If we can, then we have the percent since a percent means "over 100." For example, 9/50 can be converted to 18/100 which is just 18%.

For fractions that can't be altered like this, we must convert the fraction to a decimal like we learned by computing top divided by bottom, and then converting the decimal to a percent as described above. For example, 9/17 = 0.529 (rounded) = 52.9%

PRACTICE EXERCISES: The companion workbook offers exercises for extra practice with this topic.

INSTRUCTOR'S NOTES:

LESSON 51

AIM/TOPIC(S): Percents less than 1% and greater than 100%

WARM-UP / MOTIVATION: Elicit the student's ideas on this topic, and if/where they've seen them.

DEMONSTRATIVE EXAMPLES / POINTS TO ELICIT:
Elicit that 1% means "1 out of 100," and often we are dealing with less than that. For example, 1 out of 200 is half of 1% which could be written as 0.5%. Many students will panic and ask, "Is that a decimal or a percent?" The answer is that it is a percent which happens to contain a decimal. We can have percents involving decimals such as 38.5%, or percents that are less than 1% such as 0.42%.

Let's convert 0.42% to a decimal by following the procedure from the last lesson. The procedure was to drop the % sign and put the value over 100, giving us 0.42/100. Elicit that we used the exact value in front of the % sign. We saw that all we need to do at this point is move the decimal two places left since that is the shortcut for dividing by 100. We get 0.0042.

Percents greater than 100% are common as well, and are often found in problems (or news stories) involving percent increases. They are handled the same way. 235% is equivalent to the fraction 235/100, which in turn is equivalent to 2.35. Allow the student the sit with these examples as long as necessary.

PRACTICE EXERCISES: The companion workbook offers exercises for extra practice with this topic.

INSTRUCTOR'S NOTES:

LESSON PLANS FOR "WORKING WITH PERCENTS"

LESSON 52

AIM/TOPIC(S): Percent of increase or decrease

WARM-UP / MOTIVATION: Present this problem: A item's price increased from $20 to $30. By what percent did it increase? Elicit that the price increased by $10 which represents 50% (i.e., ½) of the original price. The price increased by 50%.

DEMONSTRATIVE EXAMPLES / POINTS TO ELICIT:
Now present this problem: During a sale, an item's price was decreased from $80 to $60. By what percent did it decrease? The price decreased by $20 which represents 25% (i.e., one-fourth) of the original price. The price decreased by 25% which would likely be promoted as a "25% Off" sale.

Elicit that the formula for computing percent increase or decrease is to take the amount of change, and divide it by the *original* price. We get a decimal which we convert to a percent.

We always compare the amount of change to the original price regardless of whether the price went up or down. For example, to compute the percent of increase from $52 to $59, compute 7 ÷ 52 to get 0.135 (rounded) or 13.5%. To compute the percent of decrease from $546 to $542, compute 4 ÷ 546 to get 0.007 (rounded) or 0.7%. Some percent changes will be less than 1% or greater than 100%.

PRACTICE EXERCISES: The companion workbook offers exercises for extra practice with this topic.

INSTRUCTOR'S NOTES:

LESSON 53

AIM/TOPIC(S): Percent "of" vs. "off"

WARM-UP / MOTIVATION: Review that the word "of" means multiplication when it appears between two values. Review the previous lesson which made reference to a price reduction during a sale offering a given percent "off" the original price.

DEMONSTRATIVE EXAMPLES / POINTS TO ELICIT:
Whenever we need to use a percent as part of computation, we should first convert it to a decimal. With this in mind, along with what was learned in Lesson 23, if we are asked to compute 23% of 594, the computation we would perform is 0.23 × 594 which equals 136.62. Have the student estimate to see if the answer is reasonable. 23% is a bit less than 25% (one-fourth), and 594 is a bit less than 600. One-fourth of 600 is 150, so our answer is reasonable.

The only point to reinforce is that the word "of" in this context translates to multiplication, and we should first convert the given percent into a decimal before using it in a computation.

There are two ways to handle problems involving a percent "off," both of which actually involve the concept of percent "of." Present this problem: "A $40 item is on sale for 30% Off. What price will you pay for the item?"

The most direct method is understanding that if an item's original price will be reduced by 30%, you will pay 70% of the original price, computed as 100% − 30%. This means we can get our answer by computing 70% of $40 (i.e., 0.70 × 40 = $28).

LESSON PLANS FOR "WORKING WITH PERCENTS"

The second method of solving this problem is to compute the amount of the discount, and then subtract that from the original price. In the above problem, we can determine the amount of the discount by computing 30% of $40 (i.e., 0.30 × 40) to get $12. Now that we know the amount of the discount, all we must do is subtract it from the original price to determine what we will actually pay. We compute $40 – $12 to get $28, which is the same answer we got using the other method.

We must be careful to see if a problem wants us to compute just the amount of discount on an item, or the price that we will pay after the discount has been subtracted.

POINTS TO REEMPHASIZE: For some reason, students tend to make this topic much more complicated than it is. Ensure that the student does not think that the words "of" and "off" are interchangeable just because they look and sound similar.

Reinforce the fact that in almost all circumstances, the word "of" can automatically be replaced with a multiplication sign. Remind the student that we should convert percents into decimals before using them in a computation, and of course review the procedure for doing so. Students tend to get confused with percents that are less than 10%, and with percents that include a decimal portion. Spend some extra time reviewing the two methods of solving problems involving percent "off," ensuring that the student is not confused by the fact that the methods themselves involve the concept of "percent of." This lesson is a very popular exam topic.

PRACTICE EXERCISES: The companion workbook offers exercises for extra practice with this topic.

INSTRUCTOR'S NOTES:

MATH MADE A BIT EASIER LESSON PLANS:
A GUIDE FOR TUTORS, PARENTS, AND HOMESCHOOLERS

LESSON 54

AIM/TOPIC(S): Increasing/decreasing a value by a given percent

WARM-UP / MOTIVATION: Ask the student for situations in which we increase a value by a given percent (e.g., rent hikes, sales tax, etc.). Then ask about situations in which we decrease a value by a given percent (e.g., sales, discounts, reductions in benefits, etc.). Use any responses to lead into the lesson.

DEMONSTRATIVE EXAMPLES / POINTS TO ELICIT:
In the last lesson we learned how to solve problems involving a percent decrease. Review that lesson as needed. We can use a similar procedure to increase a value by a given percent.

Present this problem: "A person's monthly rent of $860 is going to increase by 2.5%. What will the new monthly rent be?" There is more than one way to solve this problem, but the most straightforward is to compute the amount of increase, and then add it to the original amount. Elicit that we must compute 2.5% of $860 which we learned how to do in the last lesson: $0.025 \times 860 = \$21.50$. Add that to $860 to get $881.50. Sales tax problems are handled the exact same way.

When reading a problem, be very careful to determine if it is asking just for the amount of increase or decrease, or for the amount after the increase or decrease has been applied.

PRACTICE EXERCISES: The companion workbook offers exercises for extra practice with this topic.

INSTRUCTOR'S NOTES:

LESSON PLANS FOR "WORKING WITH PERCENTS"

LESSON 55

AIM/TOPIC(S): Two common models of word problems; Common equivalent decimals/fractions/percents

WARM-UP / MOTIVATION: Elicit the student's ideas about these problems: "6 is what percent of 30?," "What percent of 30 is 6?," and "What is 6% of 30?" Are any or all of these problems mathematically equivalent, and if so, how and why?

DEMONSTRATIVE EXAMPLES / POINTS TO ELICIT:
Let's start with the third problem. We already learned how to solve it: $0.06 \times 30 = 1.8$. This problem is not at all the same as the first two. In this problem we were given a percent, and were asked to compute that percent of a value. In the first two problems, we are not given a percent. We are given two values, and are asked to compute what percent one value is of the other. The difference may be subtle, but don't allow the student to just brush this off. Be sure to elicit that the first two problems are mathematically and linguistically equivalent.

Revisit the first problem. Again, we are not given a percent — we are asked to compute one. The format of the question is, "x is what percent of y?" We are being asked to make a comparison between these two numbers, which we do by way of division. We set up the problem as x/y which is $x \div y$. In this case we have $6 \div 30$, which is 0.2, which converts to 20%. Elicit that this answer makes sense since 6 is $\frac{1}{5}$ of 30.

Revisit the second problem. Again, this problem is mathematically and linguistically equivalent to the first one.

Lesson continued on next page →

However, if we use the same letters we did in the first problem, the format is, "What percent of y is x?" Elicit that we must set up the problem in exactly the same way we set up the first problem: x/y which is $x \div y$. This can be hard to grasp since the y appears in the problem before the x. Because of the way in which the problem is worded, we are still comparing x to y in the sense of determining what portion of y is represented by x. Ensure that the student doesn't solve this problem by computing $y \div x$ with the logic of, "y came before x."

Unfortunately, once the student manages to feel comfortable with all this, there is another scenario which can add to the confusion. Present this problem: "What percent of 8 is 12?" Almost all students will compute 8/12 to get $0.\overline{6}$ which is 66⅔%. This is wrong, and there are two ways of proving why.

First, this problem follows the model above of "What percent of y is x?" We learned that we must setup such problems as $x \div y$. In this case we get $12 \div 8$ which is 1.5 or 150%. Let's see if that makes sense. We want to know what portion of 8 is represented by 12. We're representing more than the original number itself. That means that our answer must be bigger than 100%. 12 is 4 more than 8, and 4 is 50% of 8. That means that 12 is 100% + 50%, or 150% of 12. The answer is the same for the equivalent problem of "12 is what percent of 8?"

Just like the answer to problems of this form will sometimes be greater than 100%, they will also sometimes be less than 1%. Present this problem: "3 is what percent of 600?" Elicit that it could have been equivalently worded as, "What percent of 600 is 3?" We compute $3 \div 600$ to get 0.005 which is 0.5% (i.e., half a percent). Verify that the answer makes sense. 6 would have been 1%, and 3 is half of 6.

LESSON PLANS FOR "WORKING WITH PERCENTS"

It is important to emphasize the concept that 3.5 is 50% of 7, 7 is 100% of 7, 10.5 is 150% of 7, 14 is 200% of 7, 21 is 300% of 7, and so on. This will be extremely challenging. By this point in the material, most students are mentally saturated, and are perfectly content to say, "If this is on the test, big deal, I'll just get it wrong." Unfortunately, though, percents are a favorite topic among test-makers, perhaps because they play an important role in everyday life.

There is one final important task on the topic of percents which is to gradually have the student learn and memorize the chart from the first book showing common percents and their decimal and fractional equivalents. This should not be taught by rote. As you continue to work with the student on percent problems, emphasize the common ones from the chart. Try to elicit that the percents, decimals, and fractions in the chart come up quite frequently in everyday life.

IDEAS FOR EMBELLISHMENT: See if you can get the student to take note of everywhere s/he sees percents in everyday life and in his/her other subjects, and then bring them to you for discussion. In particular, newspapers and news magazines make frequent use of percents in their articles since a great deal of news revolves around statistics. Try to convince the student that out of all the math s/he will study, the topics of fractions, decimals, and percents have the most real world relevance and practical applications.

PRACTICE EXERCISES: The companion workbook offers exercises for extra practice with this topic.

INSTRUCTOR'S NOTES:

CHAPTER ELEVEN

Lesson Plans for "Basic Probability and Statistics"

Lessons 56 to 60

Topics Covered in This Chapter:

Computing the average (mean); Median; Mean vs. median; Mode; Range; Introduction to probability; General probability formula; The chance of something not happening; Cultural aspects of probability problems; Myths and trick questions; Counting principle; Compound independent events; Probability with and without replacement

MATH MADE A BIT EASIER LESSON PLANS:
A GUIDE FOR TUTORS, PARENTS, AND HOMESCHOOLERS

LESSON 56

AIM/TOPIC(S): Computing the average (arithmetic mean)

WARM-UP / MOTIVATION: Ask the student for his/her ideas on the concept of average in math, for example, as it applies to exam scores. Elicit his/her ideas on how grades of zero affect the average. Use any replies to lead into the lesson.

DEMONSTRATIVE EXAMPLES / POINTS TO ELICIT:
The average (mean) is a value which represents the balance point of a list of numbers. For example, the average of 70 and 90 is 80. The average of 63, 65, and 67 is 65. When we have a list of numbers for which the average is not obvious, we must use a special formula to obtain it.

Most students know the formula or the general idea, so just elicit that the formula is *Sum of Scores ÷ Number of Scores*. Elicit that scores of 0 are counted, and are considered among the number of scores being averaged. (Note: You may have to address the issue of, "But the teacher drops the lowest grade.")

For practice, try computing the student's actual exam average for any of his/her classes. Also, create some "what-if" scenarios so that the student can see how higher or lower grades would have affected his/her average. In algebra we'll learn how to determine the minimum grade necessary on a future exam to ensure a minimum overall average.

PRACTICE EXERCISES: The companion workbook offers exercises for extra practice with this topic.

INSTRUCTOR'S NOTES:

LESSON PLANS FOR "BASIC PROBABILITY AND STATISTICS"

LESSON 57

AIM/TOPIC(S): Median; Mean vs. median; Mode; Range

WARM-UP / MOTIVATION: Elicit the student's ideas on what would happen to the statistic of "typical family's wealth" in a small town if a billionaire moved in. Then ask when s/he has heard the word "median" in everyday life. Offer "street" or "highway" as a hint. Use any responses to lead into the lesson.

DEMONSTRATIVE EXAMPLES / POINTS TO ELICIT:
Elicit the general idea that the mean is easily pulled up or down by extreme values. A grade of 0 easily lowers what could have been a good average, and a grade of 100 pulls it up. This is the appropriate statistic to use when it comes to exam grades. Anything else wouldn't paint an accurate picture.

Revisit the first warm-up question. If we average in the billionaire's wealth, it will greatly misrepresent the wealth of a typical citizen of the town. We have another statistic called the median which is more suited for reporting this type of data.

Revisit the second warm-up question. The word "median" means "middle." In math, the median is the middlemost value in a list after its values have been arranged in ascending order. Have the student imagine doing this for the warm-up scenario. The extreme values will not affect the median. Any very low incomes will not pull it down, and the billionaire won't pull it up. The middlemost value is a fair representation. Some test questions involve choosing whether the mean or the median is best suited for a given scenario.

Lesson continued on next page →

To compute the median of a list, we just sort the list from low to high, and then choose the middlemost value. If we do not first sort the list, we will get the wrong answer.

In a list containing an odd number of values, one value will be in the exact middle (i.e., the median). For example, the median of {14, 59, 108, 72, 999} is 72. In a list containing an even number of values, we compute the median by calculating the mean of the two middlemost entries after first sorting the list. For example, the median of {6, 28, 65, 871, 60, 4325} is 62.5.

Elicit that the median of lists containing an even number of values will only be a member of the list if the two middlemost values happen to be equal. For practice, ask the student to create some short lists of values (5 or 6 values in each), and then compute the median for each list.

Another common statistic is the mode. The mode is defined as the value in a list that occurs the most frequently. If there is one such value, then that is the one and only mode. If more than one value is tied for most-frequently occurring, then each of the tied values are considered modes. If no value occurs any more frequently than any others, then there is no mode.

The last basic statistic is the range. The range is defined as the largest value in the list minus the smallest value. If the list is unsorted, these values will not likely be the first and last in the list, so ensure that the student is being careful.

PRACTICE EXERCISES: The companion workbook offers exercises for extra practice with this topic.

INSTRUCTOR'S NOTES:

LESSON PLANS FOR "BASIC PROBABILITY AND STATISTICS"

LESSON 58

AIM/TOPIC(S): Introduction to probability; General probability formula; The chance of something not happening; Cultural aspects of probability problems; Myths and trick questions

WARM-UP / MOTIVATION: Elicit the student's ideas on what probability is all about. Assess the extent to which the student has experience with games involving aspects of probability such as coin flipping, die rolling, or card drawing.

DEMONSTRATIVE EXAMPLES / POINTS TO ELICIT:
Probability is the study of how likely it is that an event will result in a particular outcome. Informally define "event" as whatever we are doing. It may entail rolling a single die, or drawing a card followed by flipping coin. Informally define "outcome" as the result of the given event.

The odds of a coin landing on heads is "50/50" which can be expressed as either ½, 50%, or 0.5. Depending on the circumstances, a fraction, percent, or decimal might be more useful. We already learned how to convert between them.

The chance of an impossible outcome (e.g., a single die landing on 7) is 0% or 0. Elicit that the chance of a guaranteed outcome (e.g., a single die landing on 1 through 6) is 100% or 1. The probability of any outcome is always between 0% and 100% (or between 0 and 1 when converted to a decimal).

Ask the student what the chances are of rolling a 1 or a 2 on one roll of one die. Elicit that the general probability formula

Lesson continued on next page →

for an event is the number of favorable outcomes divided by the number of total outcomes. In this example we get 2/6 which reduces to 1/3. Explain that a probability problem will always define and explain the context of the scenario. The student will never be asked to compute "the probability of purple" unless all necessary details have been provided.

Elicit that if there is a 62% chance that it will rain at a given time, there is a 38% chance that it will not rain. This was computed by subtracting the given probability from 100%. If we are working with fractions or decimals, we will instead subtract from 1. If there is a ⅖ chance of a spinner landing on green, there is a ⅗ (i.e., 1 − ⅖) chance of it not landing on green. If there is a 0.25 chance of drawing a club from a deck of cards, there is a 0.75 chance (i.e., 1 − 0.25) of not drawing one.

Some students are unfamiliar with terminology and concepts commonly found in probability problems such as a standard deck of playing cards. Explain the details as needed, referring to the corresponding section in the main book for guidance.

Explain that all probability problems imply a "fair" event regardless of whether or not that word is used. There are no tricks or gimmicks or things like coins landing on their edges. Elicit that there is no such thing as a coin that is "overdue" to land on tails, nor one in which heads are on a "winning streak." Explain that many easy questions are designed to test whether the student knows better than to fall for a trick.

PRACTICE EXERCISES: The companion workbook offers exercises for extra practice with this topic.

INSTRUCTOR'S NOTES:

LESSON PLANS FOR "BASIC PROBABILITY AND STATISTICS"

LESSON 59

AIM/TOPIC(S): Counting principle; Compound indep. events

WARM-UP / MOTIVATION: Present this problem: A coin will be flipped twice. What is the chance that it will land on heads on both flips? Then present this problem: A man has 3 ties, 4 cats, 5 shirts, 6 pairs of pants. How many combinations of outfits comprised of a tie, shirt, and pair of pants can he create? Use any responses to lead into the lesson.

DEMONSTRATIVE EXAMPLES / POINTS TO ELICIT:
Let's start with the easy second-warm up question. These combination problems involve what is called the counting principle. All we do is multiply the involved numbers, being careful to not get tricked into including unrelated data. In this example we get $3 \times 5 \times 6 = 90$. The 4 cats is just decoy data.

Revisit the first warm-up problem. The coin has no memory — one flip does not affect the other. Each flip is an independent event. Many students attempt to add the two probabilities of ½ to get 1 and are then confused since it is far from certain that both flips will be heads. Elicit that when we deal with compound independent events, we multiply the individual probabilities. In this case we get ¼. Demonstrate with tree diagrams and/or sample space listings if necessary. As another example, the probability of a coin landing on heads combined with a die landing on either 5 or 6 is $\frac{1}{2} \times \frac{2}{6} = \frac{2}{12} = \frac{1}{6}$.

PRACTICE EXERCISES: The companion workbook offers exercises for extra practice with this topic.

INSTRUCTOR'S NOTES:

MATH MADE A BIT EASIER LESSON PLANS:
A GUIDE FOR TUTORS, PARENTS, AND HOMESCHOOLERS

LESSON 60

AIM/TOPIC(S): Probability with and without replacement

WARM-UP / MOTIVATION: Present this problem: An urn has 3 green marbles and 5 red marbles. One marble will be drawn, followed by another. What are the chances that both marbles will be blue? Does it matter whether or not the first drawn marble is put back before drawing the second one? Use any responses to lead into the lesson.

DEMONSTRATIVE EXAMPLES / POINTS TO ELICIT:
Revisiting the warm-up problem, elicit the concept of replacement. In any problem involving drawing multiple items (e.g., playing cards, marbles from an urn, etc.) we need to know whether or not drawn items are going to be put back (i.e., replaced) after they are drawn. Ensure the student understands that this piece of information is highly significant.

Let's first solve the warm-up problem assuming that it included the words "with replacement." That means we will draw a marble, put it back, and then draw another one. This is effectively the same as the problems from the last lesson involving independent compound events. We just multiply the individual probabilities. The probably of drawing a red marble on the first draw is 5/8. The probability of drawing a red marble on the second draw is also 5/8. We multiply those probabilities to get 25/64, and we're done.

Now let's solve the problem assuming that it included the words "without replacement." That means we will draw a marble, leave it out of the urn, and then draw another one. We will still multiply the individual probabilities, but we must carefully examine the situation after the first draw. The probably of drawing a red marble on the first draw is still 5/8. Since we're examining the odds of success, we must assume that a red marble was drawn. After that marble is gone, the urn now has 3 green marbles and 4 red ones. The probability of drawing a red marble on the second draw is 4/7. We multiply $\frac{5}{8} \times \frac{4}{7}$ to get 20/56 or 5/14, and we're done.

For practice, have the student use the same problem data to determine the chances of drawing a green marble followed by a red marble, with replacement. The answer is $\frac{3}{8} \times \frac{5}{8} = \frac{15}{64}$. Then have the student repeat this but under the condition of "without replacement." The answer is $\frac{3}{8} \times \frac{5}{7} = \frac{15}{56}$. Elicit that after the green marble was drawn and not put back, the urn still had 5 red marbles, but only 7 total.

POINTS TO REEMPHASIZE: For problems involving one event followed by another, we multiply (not add) the individual probabilities. We must always know if problems like these involve replacement. We always must consider if the first draw will change the conditions of the second draw.

NOTE TO INSTRUCTOR: Problems involving replacement at this level will all follow this pattern. In later math the student will work with problems in which we have to add probabilities that represent different possible combinations of outcomes.

PRACTICE EXERCISES: The companion workbook offers exercises for extra practice with this topic.

INSTRUCTOR'S NOTES:

CHAPTER TWELVE

How to Teach Meditation to Increase Mindfulness and Reduce Anxiety

WHY MEDITATION?

There is abundant evidence that practicing meditation can improve one's ability to be mindful, relaxed, and alert. These are three important mental conditions for succeeding with math, let alone any challenging undertaking.

Unfortunately, meditation is not part of mainstream American culture. For many people, especially young students, the thought of "just sitting there" is akin to torture, or at the very minimum a terribly silly idea and waste of time.

By practicing meditation, a student's grades will almost certainly increase. It will be easier for him/her to absorb new material, and to retain and recollect the material that s/he already knows. The student will also be more relaxed on exams, thereby not losing points due to anxiety and careless errors. The small percentage of students that practice meditation truly have a big advantage over those who do not.

BROACHING THE MATTER WITH ADULT STUDENTS

If you have an adult student who complains of or shows signs of any of the aforementioned symptoms, you can consider bringing up the idea of him/her trying to practice meditation either on his/her own or with you. It is highly recommended that any such discussion not take place during lesson time, and that you do not charge for the discussion. I also do not recommended that you charge for any mediation instruction or practice sessions which you provide for the student. If you are going to offer this special service, it really should be something that you do just for its own sake.

BROACHING THE MATTER WITH PARENTS

Do not bring up the topic of meditation with a non-adult student, even if his/her parents are present. Some parents will jump to the conclusion that you are attempting to instill your religious beliefs on the student, or that you are implying that the student has a mental, emotional, or spiritual deficiency.

If you have a non-adult student who you feel could benefit from the practice of meditation, first speak to his/her parents completely outside of the session and in private. Understand in advance that your good intentions may result in your losing the family as a client.

A COMMON RESPONSE TO THE SUGGESTION

Meditation is often misunderstood in our culture. For many, the image that comes to mind is that of a robe-clad guru sitting in the lotus position, chanting foreign language mantras in a monotone voice while incense sticks burn and gongs gong.

HOW TO TEACH MEDITATION TO INCREASE MINDFULNESS AND REDUCE ANXIETY

The stereotypical image may also involve the practitioner bowing or prostrating him/herself in front of a representation of a holy entity. For some people, the concept of meditation is synonymous with hypnosis or being in some sort of trance.

Explain that what you are proposing has nothing to do with any of that. All you will be doing is "just sitting" for a predetermined amount of time. You and the student will sit quietly in separate chairs with both of you facing the same direction to avoid self-consciousness. You're going to sit just for the sake of sitting, in an effort to bring stillness to the mind to foster a mental state that is better suited for the study of math.

The typical response to the suggestion is a simple "brush off." The client will likely say that s/he doesn't have any free time, or knows from experience that s/he just can't sit still without an activity or at least something to think about. Parents of young students may just chuckle and say, "I try all day and all night to get my ADHD child to sit still! What makes you think that you can do any better?!" As mentioned, many people will respond from the standpoint of the stereotypical mindset described above, and will have many questions about what you are suggesting.

AN ANALOGY FOR THE MENTAL RESPONSE DURING MEDITATION

If the adult student (or parent) shows any interest at all in the matter, which again is not likely, ask him/her to imagine using his/her hands to push water around in a bathtub, thereby making waves. Then ask him/her to imagine trying to stop the waves using his/her hands in some way. It isn't possible. You can't just push the waves down. In fact, if you were to try to, you would only end up making more and bigger waves. The only way to stop them is to let them settle on their own which they will quickly do if you sit still and leave them alone.

Explain that this is the goal of meditation. By "just sitting there," we first become aware off the endless and chaotic mental waves in our head. If we try to analyze them or forcibly stop them, we'll just build up more mental energy which will actually increase them. However, if we just watch them indifferently, they will gradually settle on their own, and in their own time. Certainly random thoughts will always pop up, but they will just come and go like passing clouds.

Explain to the student that by practicing meditation in this way, s/he will be able to apply the meditative state of mind in other contexts of his/her life, including studying and test-taking. The choppy ocean is transformed into a tranquil lake. The student will likely understand that such a mindset is more conducive to success than the one we usually have.

HOW TO PRACTICE MEDITATION WITH A STUDENT

If the adult student (or parent) wants to take you up on your offer, schedule a time for this purpose. An ideal time is at the end of a session since the student is already with you. Agree on how long you will try to meditate together. An initial goal might be as short as five minutes considering that many people start getting "antsy" after only a few seconds.

Make sure you have a timing device that you can easily set, and tell the student that you will be keeping track of the time so s/he won't have to watch the clock. It is ideal if your timing device does not make any type of clicking sounds, and that it signals the end of time with a pleasing, soft sound. Arrange your chair and your student's chair so that they are a few feet apart and facing the same direction, without facing anything that will be distracting.

Tell the student that you will soon start the timer. Instruct him/her to sit with his/her back straight, but not stiff. The goal is to maintain a position in which s/he can be alert, yet not uncomfortable. S/he can keep his/her eyes either open, closed, or somewhere in between. Instruct the student to focus on his/her breathing, and just let any passing thoughts come and go. Do not let the student believe that the goal is to have no thoughts at all, or to mentally chase away any thoughts that occur. Refer back to the analogy of waves in the bathtub. Just ignore them and let them settle—don't analyze them.

Before starting, tell the student that s/he should not feel as though s/he is a statue. S/he can shift positions as needed, but should make every effort to not slouch or sit in an unusual way, and should try to not fidget. Tell the student that if the experience proves to be "too much" for him/her for any reason, s/he should just do whatever s/he needs to until time is up. This may include standing up and walking away, or switching mindsets to thoughts of his/her shopping list, or anything else. S/he should not feel as though s/he is "trapped."

Start your timer and say, "Let's begin," and then start meditating along with the student. That is all there is to it. Whatever happens, happens. When the time is up, you're done. If the student has anything to say, certainly listen, but don't encourage any type of analysis session. At some later point, you can see if the student wants to try another meditation session. The most likely response is that s/he won't. The process is just too overwhelming for most people. However, if s/he wants to, you certainly should continue to offer this service, perhaps increasing the length of time of the meditation sessions.

APPLYING THE MEDITATIVE MINDSET TO STUDYING AND EXAM-TAKING

Remind the the student that s/he is developing a mental skill which can be used in times of need. If s/he is ever feeling flustered while studying, or during an exam, s/he can take a few deep breaths, and take a few moments to try to get into a relaxed and focused state of mind. Remember, meditation has nothing to do with being in a trance. It is the complete and total opposite. It is well worth taking a minute during a forty-minute exam to get oneself relaxed and alert if it will result in the remaining thirty-nine minutes actually being that way.

THE FINAL WORD

It is imperative that you have faith in your students' ability to be successful in their math goals. Students of all ages can easily see if you are not. Do not take on any client for whom you feel you cannot be of benefit. This involves learning to become honest with yourself as well as the client.

The biggest issue you will face is students who are very far behind in math, but want to catch up by taking very few sessions over a very short amount of time. It is also common for students (or their parents) to have no idea just how far behind they are in math. Remember that if a student gets high grades on exams, and gets promoted from one grade to the next, neither the student nor his/her parents have any idea that anything could possibly be wrong.

HOW TO TEACH MEDITATION TO INCREASE MINDFULNESS AND REDUCE ANXIETY

Another issue you will face is students whose goals are not aligned with their math level and ability. While I wholeheartedly believe that virtually all students can learn the math required to earn a high school diploma or GED, I don't believe that higher level math or "honors math" is for everyone.

You will need to learn to be honest with your students during consultation sessions, or privately with parents of non-adult students. If a student is very far behind in math, but only wants to take 10 sessions with you before taking a fairly high-level exam, do not smile at him/her and say, "Sure, you can do it!" That only works in Hollywood movies. Instead, discuss the situation realistically. A student who can't add $\frac{1}{2} + \frac{1}{3}$ has no business being in a second-year algebra honors class, nor should s/he be thinking about taking a graduate level exam any time soon. Be positive and encouraging, but always be realistic and honest.

Contact me via my website if you have questions about the lessons in this book, or would like to discuss the needs of a student that you are working with. Now get out there and make a difference for some students! ☺

About the Author

Larry Zafran was born and raised in Queens, NY where he tutored and taught math in public and private schools. He has a Bachelors Degree in Computer Science from Queens College where he graduated with highest honors, and has earned most of the credits toward a Masters in Secondary Math Education.

He is a dedicated student of the piano, and the leader of a large and active group of board game players which focuses on abstract strategy games from Europe.

He presently lives in Cary, NC where he works as an independent math tutor, writer, and webmaster.

Companion Website for More Help

For free support related to this or any of the author's other math books, please visit the companion website below.

www.MathWithLarry.com

Made in the USA
San Bernardino, CA
30 October 2014